CROP PHYSIOLOGY

Dr. G.C. SRIVASTAVA
Former Head
Division of Plant Physiology
Indian Agricultural Research Institute,
New Delhi-1100012

BIOTECH
2024
BIOTECH BOOKS

ISBN : 978-81-7622-228-0

Published by : **BIOTECH BOOKS**
7462-63/23, Ansari Road, Darya Ganj,
New Delhi - 110002
Phones : 011 4300 3222
Email : biotechbooks@yahoo.co.in

Laser Typesetting : **Twinkle Graphics**
Delhi - 110 088

PRINTED IN INDIA

CROP PHYSIOLOGY

A lot of information has been generated on morphological, physiological and biochemical aspects of crop plants with our own Indian crop varieties, but they are not compiled in a proper and systematic form which could be utilised by crop scientists in further improvement in yield. These information are valuable and very important. In an effort to put these information in more coherent manner on some important cereals like wheat and rice, pulse such as chickpea and oilseeds such as sunflower and rapeseed mustard, an attempt has been made in this book so that this can provide a useful information to students of crop sciences and crop scientists who can make use of these information in tailoring a crop suitable for different agro climatic conditions.

Dr. Girish Chand Srivastava born in 1946, did his Masters degree in Plant Physiology in 1965 from Banaras Hindu University, Ph.D. from the Indian Agricultural Research Institute, New Delhi in 1969. He joined Plant Biochemistry Laboratory at the University College, London to work as Post Doc Fellow under the supervision of Prof Fowden FRS from 1969 to 1971. In 1972, he joined the Indian Agricultural Research Institute, New Delhi as Plant Physiologist, Professor and Head, Division of Plant Physiology, IARI, New Delhi.

Dr. Srivastava's broad research area is Photosynthesis, Stress Physiology and Post Harvest Physiology Horticultural Crops. He published more than 140 original research papers in Indian and Foreign journals and edited several books and research journals. He has been visiting Research Scientist at the University of California at Riverside, Michigan State University and University of Minnesota. Dr.Srivastava is the recipient of several prestigious awards including R.D.Asana Medal, J.J. Chinoy award and Fellow of the Indian Society for Plant Physiology. He received the Best Teacher Award of the Indian Agricultural Research Institute, New Delhi. Dr. Srivastava has served as member of several Research advisory council, QRT of ICAR, academic councils, National and International project screening and evaluation committees, selection boards etc. He served the Indian Society for Plant Physiology as Hon. Secretary and Executive Editor.

ACKNOWLEDGEMENTS

Financial assistance given by the Department of Science and Technology, Govt of India, under USERS scheme during the course of preparation of this book is duly acknowledged.

INTRODUCTION

Green revolution in India in late sixtees played a significant role in making India a self sufficient in food production primarily due to introduction of short statured and photoinsensitive Maxican wheats which were input responsive. Since then India has moved forward to a position of real leadership in the world arena of agricultural research. However due to burgeoning population, it is estimated that India has to produce 50% more food by 2025 and this has to be achieved with limited aerable land and limited input resources.

There are already indications that yield of some important crops is leveling off whereas in other crop, there has been no significant gains. For quantum jump in crop yield potential, we must explore the possibility of further modifying the present high yielding plant types and suggests suitable cultural practices as well. In this context the role of Plant/Crop Physilogy becomes important. Several Indian scientists have suggested plant types for different crops suggesting morphological traits however there is need to include biochemical traits in defining the desirable ideotypes of different crop for different agroclimatic conditions which could better utilise light, nutrient and water.

A lot of information has been generated on morphological, physiological and biochemical aspects of crop plants with our own Indian crop varieties, but they are not compiled in a proper and systematic form which could be utlised by crop scientists in further improvement in yield. These information are valuable and very important. In an effort to put these information in a more coherent manner on some important cereals like wheat and rice, pulse such as chickpea and oilseeds such as sunflower and rapeseed mustard, an attempt has been made in this book so that this can provide a useful information to students of crop sciences and to crop scientists who can make use of these information in tailoring a crop suitable for different agro climatic conditions.

G.C. Srivastava

CONTENTS

1

WHEAT (*TRITICUM* SPP.)

Family	:	Poaceae
Sub family	:	Pooideae
Genus	:	*Triticum*
Species	:	*T. aestivum*

GENERAL INTRODUCTION

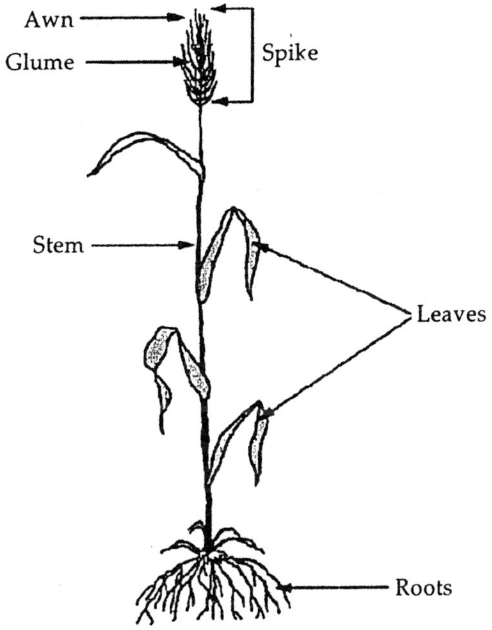

Fig. 1 : Wheat plant

Wheat (Fig. 1) is one of the oldest domesticated and worldwide cultivated crop and used as staple food in India and several other countries. Its origin is considered to be South-West Asia in the area known as the Fertile Crescent. Wild wheats were domesticated as part of the origin of agriculture in the Fertile Cresent. Although exact date of domestication of wheat may be difficult to say but it will not be wrong to say that wheats have been domesticated for about 10,000 years or little more. Wheat cultivation began to spread beyond the Fertile Crescent and reached Indian subcontinent by 6,000 cal BC. India has made significant progress in wheat production in the last few decades and ranks second amongst the largest producers of wheat after China. The green revolution in mid 60's has contributed remarkably wheat production due to introduction of semidwarf, input responsive and photo insensitive wheats. Infact what we have achieved in the quantum jump in yield is mainly because of improving the partitioning of the biomass of the irrigation and fertilizer responsive varieties and barely due to insignificant increase in biomass.

The crop is grown successfully at 25°C with minimum and maximum temperature of 3 to 4°C and 30 to 38°C respectively. Wheat is adopted to broad range of moisture conditions from xerophytic to littoral. Optimum production requires an adequate source of moisture availability during the growing season, however, too much precipitation along with high wind can cause lodging and yield losses.

Wheat grain is a staple food used to make flour for chapatti (Roti), steamed bread, cookies, cakes, pasta, breakfast cereal, noodles and for fermentation to make beer, alcohol, vodka or even biofuel.

The genetics of wheat is little complicated. Some wheat species are diploid with two sets of chromosomes (*e.g. T. monococcum*) but many are polyploids with four sets of chromosomes (tetraploid) or six sets of chromosomes (hexaploid). *T. monococcum* (Einkorn wheat) is a diploid, while most of the tetraploid wheats (emmer and durum wheat) are derived from wild emmer (*T. dicoccoides*). Hexaploid wheats are evolved in farmers' field either domesticated *emmer* or *durum* wheat hybridized with yet another wild diploid grass to make the hexaploid wheats, bread wheat.

MORPHOLOGY

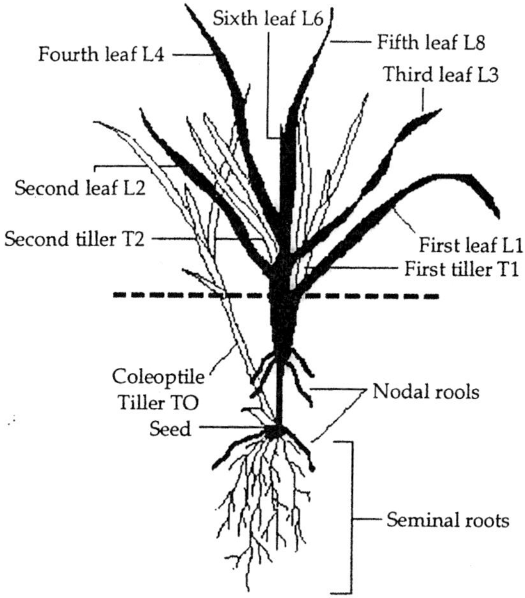

Fig. 2 : Wheat plant showing Tillers, roots etc.

Wheat stem is erect, cylindrical, hollow (in bread wheat), filled with soft pith (in Macaroni wheat), jolted, smooth and having varying plant heights depending upon the variety. Leaves have sheath, the basal part of the leaf, blade the flattened and parallel vein, ligule, the membranous out growth and auricle, the lobes of the leaf blade extending downward on both sides at the junction of blade and sheath.

Wheat produces two sets of roots during the growth period. The **seminal** or **seedling roots** are produced by the germinating seedlings at a depth where seed is planted. This root system function at the beginning of seedling growth but dies subsequently after about 30 days perhaps because of its non requirement. The second set of roots which arises from the basal nodes of the plant is called **clonal roots** later forms the compact vegetative mass better called **crown roots** and remains as permanent root system (Fig. 2). They are responsible for water and nutrient uptake besides supporting the plant physically. Wheat has inflorescence which is also called ear or head but botanically called spike (Fig. 3 and 4).

Fig. 3 : Wheat ear

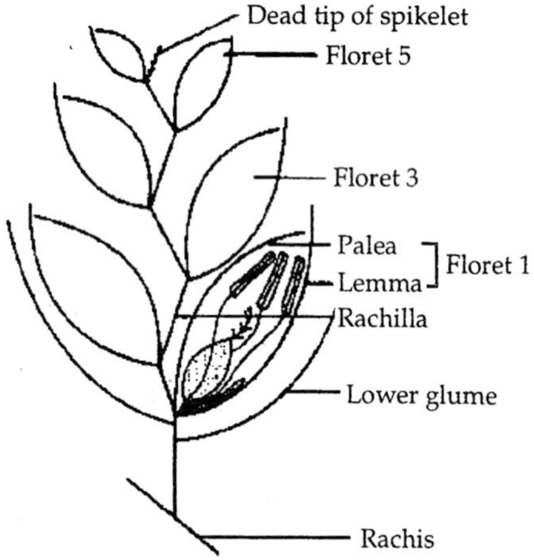

Fig. 4. Arrangement of spikelets.
*Source : **Kirby and Appleyard 1987***

This is composed of spikelets which is composed of flowers called florets. These spikelets are arranged systematically in the rachis in a zig-zag manner.

GROWTH AND DEVELOPMENT

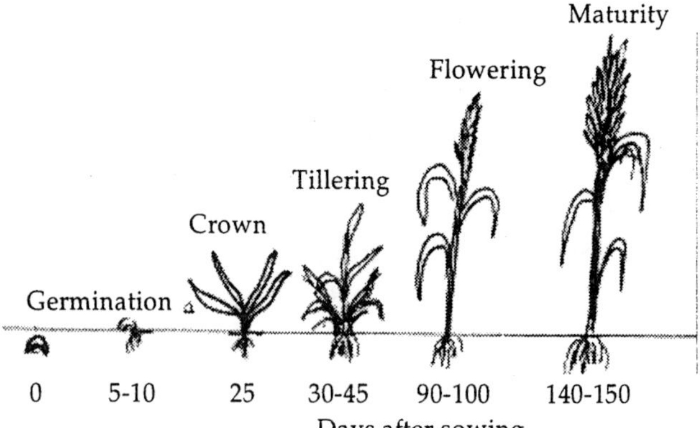

Fig. 5 : Phenological stages in wheat.

The entire grown period of wheat plant may be grouped as stage 1 (germination to emergence) stage 2 (emergence to double ridge stage) stage 3 (double ridge stage to anthesis), stage 4 (anthesis to maturity) (Fig. 5). Each growth and developmental phase/stage takes a time span which differs for different genotypes and is influenced by temperature, day length and sowing date etc.

Stage 1 (Germination to Emergence)

This is the initial phase of the entire growth period. Germination is extremely important for good growth and yield. The proper temperature and moisture level decides the rate of germination and its further growth. Several research findings have indicated that seed size does not alter the germination but does affect the growth, development and yield. Bigger seeds have advantage over smaller seed as the former has more of food reserve for initial growth. Further, it has also advantage under environmental stress condition particulars under drought (Mian and Nafziger, 1994). During germination, the seminal roots grow first followed by coleoptiles. The length of the coleoptile limits sowing depth and its length depends on the genotype. Semidwarf wheat has shorter coleoptile than the tall wheat.

Stage 2 (Emergence to Double Ridge Stage)

Fig. 6 : Double ridge stage

Wheat plant produces crown and branches out into tillers from the base at soil surface. The potential number of tillers varies with genotypes. Semidwarf types usually have high number of tillers. Differentiation of bud into tillers and tiller appearance generally ends just before stem elongation starts (Baker and Gallagher, 1983b). However, it is also suggested that tillering does not end at any specific wheat development stage, but rather is controlled by a number of genetic and environmental factors. All the tillers do not produce spike and many tillers abort before anthesis (Gallagher and Biscoe, 1978). The number of productive tillers depends on genotype and environment and is strongly influenced by density of plant.

The duration of vegetative stage in wheat depends on the sowing date and genotype and it normally varies from 60 to 150 days. It also depends on the rate of leaf appearance (phyllochrome) and time of floral differentiation which are influenced by photoperiod and temperature. Phyllochrome is defind as the interval between similar growth stages of two successive leaves in the same culm. Phyllochrome is strongly dependent on temperature but water deficit and nitrogen deficiency may retard the leaf emergence. Wheats flowers after completion of a chilling (cold) period. The double ridge stage is not reached until chilling requirements are met and the vegetative period is extended resulting in more number of leaves on the main shoot.

VERNALIZATION

Winter wheats have a strong response to vernalization and require a cold period to flowers. The vernalization requirements of winter type of wheats may be fully substituted by short days at non vernalizing temperatures. Vernalization may occur at growth stages of the growing cycle of wheat plants during germination, during vegetative plant growth and during formation of seed in the mother plant. However, the effectiveness of low temperature to accomplish vernalization decreases with increasing age of the plant.

PHOTOPERIOD

Genotypes which are sensitive to photoperiod require a certain day length to flower. However, sensitivity to photoperiod differs among genotypes. Most of the cultivated wheat are quantitative long day plants and they flower fast as the day length increases but they do not require a particular length of day to induce flowering (Major and Kiiry, 1991). It is the mature leaves which sense the photoperiod and not the apical meristems (Bernier *et al.*, 1993). A single leaf is usually enough for sensing photoperiod and to cause floral induction. Once the photoperiod insensitive period is over, floral induction starts and the reproductive stage (double ridge) begins. Short day length results in longer inductive phase and longer the phyllochrome, bigger the flag leaf.

The inflorescence development also depends on the day length in the photosensitive genotypes. Shorter the days, longer the double ridge to terminal spikelet period. Floret initiation and anthesis is however not affected by the change in day length.

Stage 3 (Double Ridge to Anthesis)

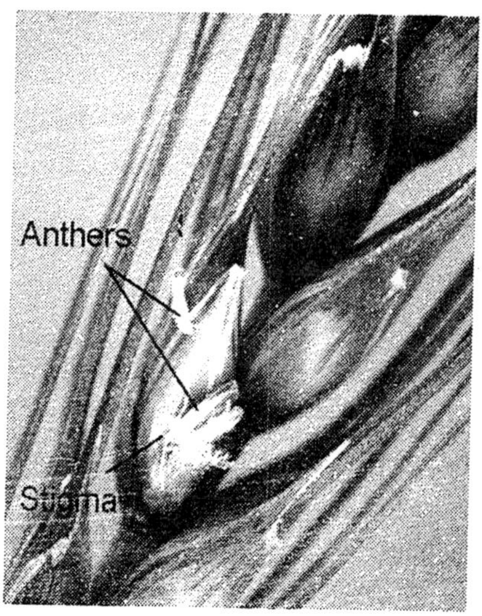

Fig. 7 : Anthesis stage

At the time when growing apex changes from vegetative to reproductive stage (Fig. 7), wheat plants have 4 to 8 leaves on the mains shoot. The floral primordia are found in the axil of each lamina. Temperature above 30°C during floret formation cause complete sterility (Saini and Aspinall, 1982).

Each spikelet has 8 to 12 floret primordial in the central part of the spike. Basal and distal spikelets have 6 to 8 florets. More than half of these florets abort or insufficiently developed before anthesis (Hay and Kirby, 1991).

Rahman *et al.* (1977) have reported a positive correlation between length of the vegetative phase and the number of spikelet per spike and lengthening the duration of the vegetative stage of the apex induces more spikelets per spike. However, the actual number of spikelets is determined by the length of the reproductive phase.

Once the terminal spikelets is formed, stem elongation starts and the spike begins to grow. Spike growth occurs from the appearance of the flag leaf upto 10 days post anthesis (Kirby and Appleyard, 1984). Higher radiation increases the amount of photosynthates available for spike growth and lower temperature prolongs the period of spikelet growth and reduces in competition for assimilates.

Stage 4 (Anthesis to Physiological Maturity)

Wheat has only one spikelet per rachis and each spikelet has 3 to 6 potentially fertile florets which are self pollinated. Anthesis begins in the central part of the spike and moves towards basal and apical parts during 3 to 5 days period (Peterson, 1965). The proximal florets of the central spiket are fertilized 2 to 4 days earlier than the distal florets and thus these grains are heavier (Sunmons and Crookston, 1979).

After the floret fertilization, cell division is fast and during this period endosperm and amyloplasts are formed. This is called lag phase where 30 per cent grain filling period occurs. Thereafter cell growth, differentiation and starch deposition in the endosperm takes place which corresponds to linear grain growth (50-70 per cent of grain filling period). Embryos is formed at the time of endosperm growth.

GROWTH RATE

Growth analysis is very important and useful method to analyze the plant/crop potential. There are various indices which are normally followed in growth analysis.

Net assimilation rate is the increase in dry matter per unit leaf area per unit time and this can be calculated following the equation.

$$\text{NAR} = \frac{(W_2 - W_1)(\text{Log}_e L_2 - \text{Log}_e L_1)}{(t_2 - t_1)(L_2 - L_1)}$$

Where W_2 and W_1 and L_2 and L_1 represent the total plant dry weight and leaf area at time t_1 and t_2 respectively (Watson, 1952).

Relative Growth Rate (RGR) : Originally the RGR was termed as the efficiency index of dry matter production (Blackman 1919) but today many still believe that the relative growth rate is the most important index of productivity (Radosevich and Osteryong, 1987). RGR is the amount of growing material per unit dry weight of plant per unit time and is calculated following the equation.

$$\text{RGR} = \frac{\text{Log}_e W_2 - \text{Log}_e W_1}{t_2 - t_1}$$

Where W_1 and W_2 are the total dry weight at time t_1 and t_2 respectively. It can also be expressed as kg day^{-1}.

The relative growth rate serves as a measure of dry matter production and can be used to compare the performance of species or the effect of treatment under defined condition. Crop growth rate is the multiplication product of leaf area ratio and leaf area index.

$$\text{CGR} = \text{LAR} \times \text{LAI}$$

In serves as a simple index of crop product units. Studies of an old and modern wheats of Australia has revealed that relative to old cultivars, modern wheats are characterized by a greater RGR during the vegetative phase. This has achieved at a lower initial green area index which persisted as the season progressed and was associated with a higher CGR at anthesis and greater grain yield at end of the season. In old cultivars a greater green area index during mid season declined rapidly as temperature and soil moisture stress increased in spring resulting in a lower green area index at anthesis. Together with lower CGR at anthesis this resulted in less dry matter and grain yield at harvest. The higher grain yield of modern wheat cultivars was achieved with higher RGR during the vegetative phase and greater CGR from ear emergence to harvest (Karim and Siddique, 1991).

Constraints in improving yield of wheat has been analyzed at various levels by a number of workers in India and outside the country. Significant improvement in productivity of wheat throughout the world has been possible by development of semidwarf lodging resistant high yielding varieties (Jain, 1986, Chandler, 1972). The enhancement in yield of the semidwarf wheat has been achieved by improving the harvest index without increasing the biomass (Jain 1986). However, harvest index has now approached a ceiling value and further increase in productivity is not possible with the present plant type unless the biomass in the present types is enhanced. This requires an effort to increase the biomass through enhancing the photosynthesis, the only process of adding more carbon to develop a new plant type.

PHOTOSYNTHESIS

Photosynthesis is the only process which add carbon to the total biomass of the plant. Wheat is a C_3 plant and its carbon fixation rate normally varies between 14 to 20 mg CO_2 dm^{-2} h^{-1}. However, this may vary due to temperature. Temperature enhances the rate of photorespiration as well as dark respiration. The rising level of CO_2 is expected to enhance the rate of CO_2 fixation by suppressing photorespiration. However, the increase in CO_2 if accompanied with rising atmospheric temperature, the net increase in photosynthesis may not be achieved as expected.

Photosynthesis varies also with age of the leaf and its position on the stem and time of the day. Young developing leaves require more of carbon for its own growth and become a sink for a while. Larger leaves normally have low photosynthetic rate compared to smaller leaves as in case of diploid wheats (Dustone *et al.*, 1973). Austin *et al.* (1982) have studied the rate of net photosynthesis in 15 genotypes of wheat and related species. At the stage when rates were maximal, they were in general highest for the diploid, species, intermediate for the tetraploid species and lowest for *Triticum aestivum* means of 38, 32 and 28 mg CO_2 dm^{-2} h^{-1}, respectively (Table 1).

Rates were strongly negatively correlated with the leaf area, width and the mean plant area per mesophyll cell and positively correlated with a stomatal frequency and number of veins per mm of leaf width. The differences among species in these attributes were mainly related to ploidy level. Current information on polyploidy and photosynthesis in several species has been discussed with respect to anatomy, biochemistry and bases for expressing photosynthetic rates. Since high yielding ability is strongly associated with larger leaves, it may display a negative relation with photosynthetic rate (Planchon, 1969, Stoy, 1965). Natr (1966) found a positive relation of photosynthetic rate with yield but only late in the season when rates were falling. The results are more of an expression of the fact that leaves in the high yielding cultivars tend to senesce later. Some genotypes however have both higher photosynthetic rate and large leaves (Craver and Nero, 1990, Morgan and Lecain, 1991) and they might become very useful in breeding programme for higher productivity.

Being a C_3 plant, the CO_2 fixation in wheat is catalatized by an enzyme Rubisco which has been considered as a potential target for manipulation for reducing photorespiration. This enzyme requires high Km for CO_2 and is likely to have photorespiration under normal condition when O_2 is present. The site for CO_2 and O_2 is same and thus a competition. Efforts have been made to modify the Rubisco to have more affinity for carbon dioxide than oxygen. Such efforts have been made *in vitro* using some chemicals like urea to alter the ratio of two activities (carboxylase and oxygenase) which showed possibility of changing this ratio however still not much success has been achieved (Ghildiyal, 1988).

Table 1 : Maximum Rates of Photosynthesis and Stomatal Conductances of Flag Leaves of *Triticum* and *Aegilops* Species

Genotype and genome	*Net photosynthesis rate per unit leaf area (mg CO_2 m^{-2} h^{-1})*	*Net photosynthesis rate per unit chlorophyll content (mg) CO_2 mg^{-1} h^{-1})*	*Stomatal conductance to water vapour (cm s^{-1})*	
			Abaxial surface	*Abaxial surface*
Hexaploids (AABBDD)				
T. aestivum L.				
Maris Huntsman	26	5.8	0.19	0.51
Hobbit	27	5.6	0.30	0.62
Professeur Marchal	26	5.5	0.19	0.53
1714/H	31	6.1	0.21	0.58
1714/L	30	5.7	0.20	0.61
Mean	28	5.7	0.22	0.57
Tetraploids (all AABB)				
T. durum Desf. x *turgidum* L.	28	5.4	0.34	0.62
T. dicoccum Schubl.	32	7.1	0.23	0.46
T. dicoccoides Korn.	35	6.2	0.31	0.57
Mean	32	6.2	0.29	0.55
Diploids				
T. monococcum L (AA)	30	11.2	0.27	0.45
T. thaoudar Reut. ex Boiss (AA)	42	7.2	0.36	0.56
T. urartu Tum. (AA)	43	9.7	0.33	0.58
T. aegilopoides Bal. (AA)	34	8.5	0.21	0.44
A. squarrosa L. (DD)	39	9.4	0.27	0.46
A. speltoides Tausch 'A' (BB)	39	7.8	0.35	0.57
A. speltoides Tausch 'D' (BB)	37	7.6	0.41	0.59
Mean	38	8.8	0.31	0.52
Genotype LSD (P=0.05)	3.2	0.9	0.06	0.07

Source : Austin *et al.* (1982)

A wide screnning of the photosynthesizing organisms have shown some hope to achieve higher photosynthetic activity. As reported by Uemura *et al.* (1977), red algae has 2-3 times more efficient rubisco than higher plants. Similar reports are there in other organism (Read and Tabita, 1994). This offers a good scope to enhance photosynthetic efficiency in higher plants through genetic transformation. Rubisco in plant normally exist in inactive form and for activity it is activated by CO_2, the substrate or by Rubisco activase enzyme. Attention have also been given on Rubisco activase in order to enhance photosynthesis through Rubisco enzyme. This activase is very vulnerable to higher temperature (Crafts-Brandner and Salvucci, 2000).

PHOTORESPIRATION

Being a C_3 plant, wheat loses a significant amount of carbon in photorespiration which is considered a wasteful process. The existence and importance of photorespiration was first clearly described by Decker and Tio (1959) based on the fully reproducible post illumination CO_2 burst they observed in many C_3 species. The burst decelerated for several minutes following darkening after a period of photosynthesis.

Table 2 : Rates of Photorespiration in Leaves of Various Species

Species	*Method of Assay*	*Temperature (°C)*	*Photorespiration, % of net photosynthesis in normal air*	*Reference*
Soybean	CO_2 release, CO_2-free air	26	46	Samish *et al.* (1972)
Tobacco	Post-illumination CO_2 burst	25.5	45	Decker (1957)
Wheat	Net CO_2 uptake 21% O_2 vs 2% O_2	28	53	Keys *et al.* (1977)
Wheat	CO_2 release, CO_2-free air	28	69	Keys *et al.* (1977)
Wheat	Short-time uptake, $^{14}CO_2$ minus $^{12}CO_2$	28	17	Keys *et al.* (1977)
Maize	CO_2 release, CO_2-free air	35	0	Moss (1966)

Control of photorespiration has emerged as a primary objective in efforts to increase plant productivity in those crops possessing the C_3 pathway of photosynthesis. Recent progress in characterizing the photorespiration and its regulation has identified the target site for modifying this process in order to increase yield. There is variation in photorespiration and photosynthesis at species and variety level. (Table 2). In tobacco, low photorespiration was correlated with high photosynthetic efficiency (Zelitch and Day, 1968). In wheat, Holbrook *et al.* (1984) have reported small and insignificant variation in photorespiration at species level. In extensive studies, a comparison of relative rates of RuBP carboxylase to oxygenation between species was made by Jordan and Ogen (1981). Holbrook *et al.* (1984) reported that within the species of *Triticum* and among other particular genotypes studied, the variation in sensitivity to O_2 is within the range of random error of present assay procedure. The variation in sensitivity to oxygen of carboxylase from various species of higher plant was also shown to be small by less direct method (Bird *et al.*, 1982).

Several efforts have been made to reduce the rate of photorespiration using some chemical inhibitors but not significant results have been obtained on yield. It is also worth considering that photorespiration since it occurs in C_3 plants only, must have some positive role as well. Our understanding regarding the role of photorespiration in C_3 plants is still not revealed clearly. However, in the recent past, a number of workers have shown that photorespiration protects C_3 plants from photooxidation and suppression of photorespiration causes severe photoinhibition (Akiko Kozaki and GO Takeba, 1996). Photorespiration is effective in dissipating excess photochemical energy. It has also been shown that a variety of photorespiratory mutants of the C_3 plant *Arabidopsis thaliana* are able to grow in a high CO_2 environment in which plants do not need to photorespire, but such plants can not grow in normal air in which photorespiration can occur (Ogren WL, 1994a , Somerville, 1986) and therefore photorespiration is necessary for survival of the C_3 plants in normal air.

Extensive efforts to increase the specificity of Rubisco for CO_2 relative to O_2 and thereby increase the productivity of C_3 crops have proved unsuccessful. In one experiment with *Arabidopsis* and wheat (both C_3 plants), Rachmilevitch *et al.* (2004) have shown that by inhibiting photorespiration under high CO_2 condition or low oxygen concentration there is strong inhibition of nitrate assimilation and this raises concern about genetic manipulations to diminish photorespiration in crops (Fig. 8).

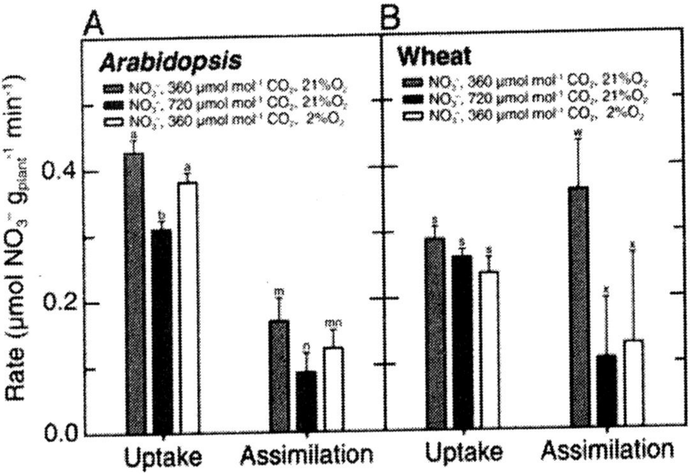

Fig. 8 : In wild-type *Arabidopsis* and wheat, NO_3^- uptake as the amount of NO_3^- depleted from a medium and NO_3^- assimilation as the difference between the rates of net NO_3^- uptake and net accumulation of free NO_3^- in plant tissues. Thirty-six-d-old *Arabidopsis* plants (A) or 10-d-old wheat (B) were exposed to either 360 µmol mol^{-1} CO_2 and 21% O_2 (gray), 720 µmol mol^{-1} CO_2 and 21% O_2 (black), or 360 µmol mol^{-1} CO_2 and 2% O_2 (white). Shown are the mean ± SE (n=13-16). Treatments labeled with different letters differ significantly ($P \leq 0.05$). The light levels were 500 and 1000 µmol $m^{-2}s^{-1}$ PAR for *Arabidopsis* and wheat, respectively.

Source : Rachmileviteh *et al.* (2004).

RESPIRATION

Respiration is most important plant process for the energy production and survival of any living organism. Respiration has a profound influence on crop yield because dry matter accumulation by a crop is closely related to CO_2 assimilation and respiratory activity. This process reduces the total built up dry material and thus affects the total dry matter accumulation. However, it is necessary for the syntheses of various vital plant metabolites. Too much respiration is also not a desirable characters. Respiration in leaf is affected by age of the leaf, temperature, season and also the cultivars (Swada, 1970). It also varies depending upon the rate and stage of growth. Dwarf varieties of wheat has much higher rate of respiration than medium height cultivars (Rawson and Evans, 1971). There are considerable difference amongst the cultivars in loss of stem weight by respiration during grain filling. In general, these accounts for about one third of the total loss in stem weight. The remainder presumably being due to translocation. Dark respiration is not proportional to the dry matter accumulation or the leaf area index but it follows a plateau when leaf area index is very high (around six).

McCree (1974) demonstrated that the respiration ® of crop can be separated theoretically into two components, growth respiration (Rg) and maintenance respiration (Rm) Growth respiration is proportional to the total assimilation (P) and maintenance respiration is proportional to dry mass (W).

$$R = Rg + Rm = KP + CW$$

Where K is the coefficient for growth respiration and C is the coefficient for maintenance respiration. The growth respiration is the cost of converting the immediate product of photosynthesis into plant material. The maintenance respiration refers to CO_2 that results from protein breakdown and the CO_2 released in respiration that provides energy for maintenance. The maintenance process includes cellular structure, and gradients of ions and metabolites, and also the process of physiological adaptation that maintain cell and active unit in changing environment. Maintenance respiration has great influence on wheat yield as dry matter accumulation and partitioning into the grain yield by the crop are closely related to assimilation of CO_2 and respiratory activity. There is cultivar variation in respiratory rate for many crops. Much depends on the stage of the plant growth at which comparison is made. In young plants when synthesis respiration is dominant component, it may be expected a positive relationship but at later stage when maintenance respiration is dominant, the relation between growth and respiration may be negative.

Global climate change causing elevated CO_2 level and temperature is likely to affect the respiration rate. Reuveni and Bighee (1997) from their experiment on high CO_2 effect studies have shown that there is considerable reduction in respiration due to high CO_2 concentration and thus seed set in wheat growing under very high CO_2 level was due to reduction in respiration rather than photosynthesis since most of the respiration was functional and needed for seed development. The rate of respiration usually doubles with every 10°C rise over the range of 5-35°C temperature (Table 3) (J.F. Talling, 1961). Friend *et al.* (1962) found dry matter accumulation in vegetative wheat (*T. aestivum*) plants to be maximal at 10°C and to drop to 50% of that value at 25°C. Flag leaf senescence was accelerated when plants were grown at higher temperature.

Table 3 : Effect of Temperature on Dark Respiration of Wheat Cultivars as Percent of Rate at 35°C

	Temperature (°C)							*Respiration rate at 35°C (mg CO_2 hr^{-1} dry wt^{-1})*
Cultivar	*5*	*10*	*15*	*20*	*25*	*30*	*35*	
Reproductive								
Scoutland	6	17	24	42	58	77	100	9.8
Ponca	6	9	21	33	47	74	100	7.6
Vegetative								
Scoutland	19	29	36	49	58	77	100	11.2
KanKing	16	31	44	52	64	84	100	11.1

Source : Todd, (1982)

Temperature optimum for the flag leaf of wheat was 20°C while that of spike was 10°C. Respiration rate increased with increasing temperature giving a Q_{10} of between 1.4 to 2.3 for the range of 15-35°C for 5 to 15°C the Q_{10} was between 1.9 to 3.7 (Table 4).

Table 4 : Calculated Q_{10} Values for Dark Respiration of Barley and Wheat Shoots (Numbers Following Averages are Standard Deviations)

	Q_{10} temperature range (°C)		
	15/5	*25/15*	*35/25*
Wheat (2 cultivars)			
Vegetative	1.89 ± 0.07	1.37 ± 0.21	1.98 ± 0.08
Reproductive	3.74 ± 0.68	2.28 ± 0.17	2.04 ± 0.29
Barley (3 cultivars)			
Reproductive	2.03 ± 0.39	2.04 ± 0.80	1.76 ± 0.29

Source : Todd (1982)

A significant amount of respiration takes place during the reproductive phase of the wheat plant, noticeably by ear (Table 4). Developing grains may contribute 60% or even more of ear respiration (Evans and Rawson, 1970) and it increases along with grain filling due to higher metabolic activities. Since initially the grain is green and capable of CO_2 fixation, the generated CO_2 in respiration is reassimilated by grain themselves as sufficient light for photosynthesis is transmitted through the glumes (Evans and Rawson, 1970). Glumes thus are very advantageous so far as photosynthesis is concerned.

SOURCE AND SINK RELATIONSHIP

Source in its broadest definition could be considered as a place from which a required assimilate is obtained whereas sink constitutes the organs receiving these assimilates. In plants, supply comes from photosynthesizing organs (leaf, bract, fruiting structure etc.) designates as source while demand is created in non photosynthesizing organs (non green plant part, like stem, root, fruit, seed etc.) termed as sink which utilize the assimilate for their growth or storage purpose. Vascular tissue (especially phloem) is the main channel for assimilate flow.

Sucrose, a product of photosynthesis and the main form in which assimilates are translocated out of leaf (Shiroya, 1977) cause feedback inhibition of photosynthesis in event of its accumulation in the leaves due to reduction in sink demand (King *et al.*, 1967, Neales and Incoll, 1968).

Judicious balance between demand and supply of assimilate is a kay to achieving higher productivity in crops where economic yield constitutes a fraction of the biomass. Opinion however, differs on whether economic yield is enhanced through efficient partitioning and utilization of available assimilates or further improvement in net photosynthesis is required to fulfill the desired goal. It is likely that the relationship varies in different crops.

Analysis of yield in wheat over eighty years, has shown that whatever increase in yield in wheat has been achieved is only by increasing the partitioning of the assimilates (Kulshrestha and Jain, 1982, Fig. 9).

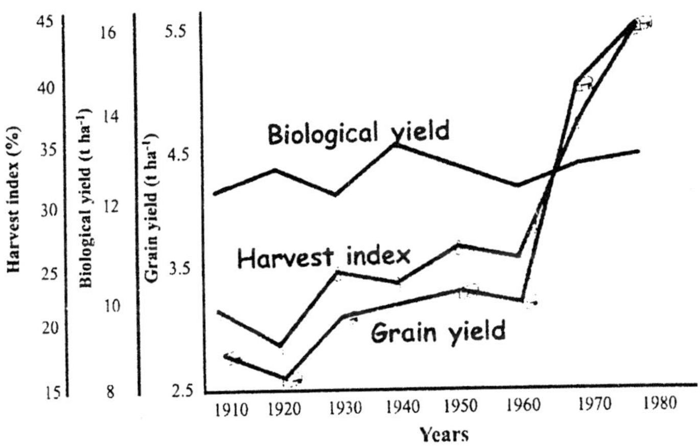

Fig. 9 : Trend in varietal yield potential of wheats in India over 80 years compared with changes in their biological yield and harvest index.

Source : Kulshrestha and Jain (1982)

Studies conducted on a large number of wheat genotypes have shown that higher rate of photosynthesis is not necessarily associated with higher grain yield since grain yield is dependent not only on the availability of assimilates but also on the ability of developing grains to accumulate and utilize them for the synthesis of reserves. The grain yield obtained may or may not necessary reflect the growth potential of the developing grain. Studies revealed that diploid wheats are more efficient in photosynthesis than tetraploid or hexaploid wheat but their yields are rather poor. It also appears from the studies that diploid wheats have lesser capacity to synthesise reserve than tetraploid or hexaploid wheats indicating that poor grain growth of diploid wheats due to its poor sink capacity (Ghildiyal and Sirohi, 1986). Diploid wheats possessed a relatively high number of spikelets (per ear) and tillered profusely. Photosynthetic rate was quite higher in commensuration with the demand of the sink (*i.e.* developing spikelets). Photosynthetic rate however decreased with diminished sink demand during post anthesis owing to poor grain growth and its capacity to accumulate reserve material. Sugar accumulate in the leaves (as mentioned earlier) decreases photosynthesis due to limitation of Pi required for RuBP regeneration. Rubisco is activated by Rubisco activase (Salvucii *et al.*, 1985) which is energy driven process requires ATP. A decrease in ATP/ADP ratio when Pi is limited causes decreased Rubisco activation resulting in reduction in photosynthesis.

Reduction in Pn due to feedback inhibition could also be removed by improving sink. In many of the crops, there are limitation in the sink potential which is prone to feedback inhibition of photosynthesis. In wheat the most important part of sink is the ear which requires assimilate for grain number and its growth. In an effort to enhance the sink capacity at a time, branched ear was suggested. A branched ear obviously should have a larger sink capacity. The semidwarf variety, Kalyansona, was compared with a branched-ear culture (from crosses of *Triticum turgidum, Triticum durum* Desf, Sonora 64 and Kalyansona). Both Kalyansona and the branched ear culture gave a similar biological yield under pot-culture. The grain yield of Kalyansona was, however, considerably higher apparently because of its high grain number per ear (85 against 25 of the branched ear). Dry weight changes, after anthesis, showed that more dry matter accumulated in the ear of Kalyansona in spite of smaller leaf area (Salunke and Asana, 1971). This idea was not found useful. A long ear is the alternative to achieve better sink capacity. Efforts have been made and success has also been achieved to some extent in getting long ear wheat which has low negative relationship of grain number with its weight. However, there is need to develop a plant with better sink capacity to enhance the biomass and economic yield as well.

WATER REQUIREMENT/WATER USE EFFICIENCY (WUE)

Water requirement by definition is the amount of water transpirated (evapotranspiration) to produce a unit of dry matter.

$$\text{Water requirement} = \frac{\text{Evapotranspiration}}{\text{Dry matter production}}$$

Water use efficient is the reciprocal of the water requirement

$$\text{WUE} = \frac{\text{Dry matter production}}{\text{Evapotranspiration}}$$

Water use efficiency refers to yield in relation to water used to produce the yield. (Zaffaroni, E. and Schneiter, A.A. 1989).

There is large difference in WUE in C_3 and C_4 plants since the amount of CO_2 fixed by C_4 is almost double the C_3 plant.

Table 5 : Water Use Efficiency [g dm. (kg $H_2O)^{-1}$] for C_3 and C_4 Species

Species	*Grasses*	*Dicots*
C_3	1.49	1.59
C_4	3.14	3.44

Low water use efficiency of crop is a wide spread feature of dry environment. There is pressing need to improve the water use efficiency of rain fed and irrigated crop production. Breeding crop varieties with higher water use efficiency is seen as providing part of solution. The water use efficiency can be improved by following some key processes such as (i) moving more of the available water through the crop rather than it being wasted as evapotranspiration from the soil surface, (ii) enhancing carbon assimilation (biomass) in exchange of water transpired and improving crop transpiration efficiency and (iii) improving harvest index. These three processes are however not independent. Targeting one specific trait may have detrimental effect on the other but there may also be possible interactions. For C_3 species, increasing carbon isotope discrimination provide a powerful means of improving water use efficiency of leaf gas exchange however, it has also been observed that improvement in leaf level water use efficient many not always translate into higher crop water use efficiency or yield.

In wheat using carbon isotope discrimination technique, a significant genotypic variation in WUE (ratio of grain yield to water used), evapotranspiration efficiency and harvest index has been reported (Ehdaie, 1995). Modern short cultivars on average had higher WUE that old tall cultivars in well watered (1.225 vs 0.984 g kg^{-1}) and droughted (1.198 vs 0.860 g kg^{-1}) pot condition due mainly to higher harvest index (Ehadaie, 1995).

NITROGEN UPTAKE AND ASSIMILATION

Nitrogen amongst the plant nutrients is the most important element required for plant growth and yield. About 2% of the dry weight in plant is nitrogen. Nitrogen deficiency in soil is a worldwide problem. Excess application of nitrogen is an environmental hazard. Scheduling nitrogen application for maximum efficiency is most desirable for judicious use of the fertilizers and to improve the yield. Improving the timing of nitrogen application will be beneficial in three ways :

1. It can improve crop yield and quality.
2. It can reduce fertilizer cost.
3. It can help protect the environment.

Split N fertilizer application often increase the efficiency of crop N use. Like split N application, slow release N applications can be beneficial in situation where a higher potential for nitrate loss from the root zone exists. For most crop, late N fertilizer application reduce the crop quality. Some examples of the determental effect of excessive N late in the season include high protein in soft white winter wheat and high sugar content and dark fry colour in potato. In addition, N remaining in soil after harvest is vulnerable to loss by leaching. Excess soil N after harvest is a potential pollutant of ground water.

The uptake of nitrogen mostly in the form for nitrate by root cell is a key step in nitrogen metabolism and has been widely studied at the physiological and molecular levels. Understanding how nitrate uptake is regulated will help us engineer crops with improved nitrate uptake efficiency. Woodland *et al.* (1986) while studying the uptake and utilization efficiency of nitrogen in tall, semidwarf and dwarf Indian and Mexican wheats reported that tall varieties were efficient in nitrogen uptake and utilization over dwarf. Dwarf varieties were good in nitrogen utilization efficiency for grain yield. Similar observation has been made by Singh and Arora (2001) in wheat plants (Table 6).

Table 6 : Comparison of Nitrogen Uptake and Utilization Efficiency in Two Groups of Wheat at two Levels of Nitrogen

Genotypes	*Nitrogen Uptake* (N_P *g pot*$^{-1}$)		*NE1*[1] (*g g*$^{-1}$)		*NE2*[2] (*g g*$^{-1}$)	
	40 kg N	*120 kg N*	*40 kg N*	*120 kg N*	*40 kg N*	*120 kg N*
Dwarf (mean	0.724	1.026	117.4	102.6	45.4	38.4
Tall (mean	0.806	1.086	127.6	109.2	39.5	33.6
CD at 5%						
Groups	0.061	0.043	7.07	3.73	2.55	3.15
Nitrogen levels	0.21	18.29	6.79			

[1]NE1 = nitrogen utilization efficiency for biomass production; [2]NE2 = nitrogen utilization efficiency for grain production.

Source : Singh and Arora (2001)

The uptake of nitrate involves nitrate transport systems. Two transport systems for nitrate have been identified based on kinetic studies i.e. high affinity transport system (HATS) which has high affinity for nitrate and works at low concentration of NO_3 in the media is constitutive as well as inducible. Low affinity transport system (LATS) has low affinity for NO_3 and works at high concentration of NO_3 in the media and is constitutive in nature. Both the transport systems work depending on the concentration of NO_3 in the media. Efforts are being made to study the regulation of high affinity transport system to improve the nitrate uptake. In wheat, Cai *et al.* (2007) have shown the role of exogenous abscisic acid (ABA) and glutamine (Gln) in regulation of genes. The result in the field with improved nitrate uptake has yet to be seen. The genes coding for LATS and HATS have been identified in higher plants (Daniel-Vedele *et al.* 1998). The over expression and under expression of these genes is of great significance in developing transgenic for improved nitrate uptake.

Nitrate once taken by the plant is converted to organic form of nitrogen. Nitrate assimilation is essentially an enzymatic process. It is reduced to nitrate by enzyme nitrate reductase (NR) to nitrite which is subsequently reduced to ammonia by enzyme nitrite reductase and further metabolism proceeds by way of other enzymatic reactions in which nitrogen is incorporated into a wide range of nitrogenous organic compounds. Enzyme nitrate reductase is considered as a rate limiting enzyme. Beevers and Hageman (1969) considered that under conditions of non limiting soil nitrate, plants with higher levels of NR activity should posses a greater potential for accumulating reduced nitrogen. The enzyme is highly heritable. There is large variation in the activity of nitrate reductase *in vivo* and *in vitro* in wheat genotypes when expressed on a dry weight basis (Abdin *et al.*, 1992). The genotypes with high nitrate reductase (HNR) activity exhibited an efficient nitrate induction system. Further studies have shown that HNR genotypes had more active enzyme protein, high rate of enzyme protein synthesis and higher level of steady state nitrate reductase-mRNA relative to those observed in low nitrate reductase (LNR) genotypes (Abdin *et al.*, 1992). It may be mentioned that over expression of either NR or the NiR genes in plant increases mRNA level and often affect N uptake. However, the increase uptake of N does not seem to increase the yield or growth of plants regardless of the N source (Andrews *et al.*, 2004). Improvement in NUE via manipulation of further processes like plastidic glutamine synthetase-2 (GS-2) and Fd-glutamine oxoglutarate aminotransferase (Fd GOGAT) genes has also not provided significant results.

Studies with transgenic plants over expressing genes affecting the nitrogen metabolic pathway suggest that it is possible to improve or manipulate N metabolism and the growth phenotype of plants which can improvė the NUE of crop plants. However, inspite of several studies over the past few years both at whole plant level and using transgenic plants, our understanding of the uptake and utilization of nitrogen is still at not clear and requires further research.

ABIOTIC STRESS

The productivity of plant is greatly affected by abiotic stresses such as water stress, temperature, stress, salt and heavy metals stress etc.

WATER STRESS

Amongst the abiotic factors affecting wheat productivity, drought is the single most important factor severely affecting the yield. There is conscious effects for improving the wheat productivity under the abiotic stress environment and therefore efforts are being made to understand the mechanism of tolerance at whole plant level as well as at molecular level. The severity of stress effect on crop depends on both intensity and the duration of the drought.

Stress affects is visualized right from germination to grain filling stage. Moisture is required for germination of the wheat seed and shortage of moisture at this stage causes delay in germination and even death of the seed. Good germination and seed vigour are very important in achieving a good crop.

Considerable genotypic variation has been reported for seed germination, root/shoot length, coleoptile length, osmotic membrane stability in wheat under normal condition than under moisture stress indicating the suppression of variability under moisture stress condition. Comparing mean performance under normal and osmotic stress condition, Dhanda *et al.* (2004) reported that seed vigour index is the most sensitive trait followed by shoot length, germination percentage and root length. The pre soaked seeds in water by virtue of quicker and better germination under soil moisture stress condition are able to establish better and consequently yields higher. Pre-hardening of seed has been reported to show beneficial effect. As early as 1942, Chinoy reported that wheat plants grown from pre-hardened seeds reached anthesis earlier than those grown from untreated seeds and the earliness produced by the pre-sowing treatment was highly significant. Chinoy *et al.* (1970) further reported higher grain yield of wheat in pretreated seed with ascorbic acid. The pretreated seeds gained an appreciable advantage over the control in the seedling stage. This advantage was brought out by acceleration in dry matter production as well as relative growth rate not only under normal water availability but also under restricted water supply, drought and salinity condition.

The effect of water stress at different stages is not uniform (Chinoy, 1960, 1961, 1962). Investigation on the water stress effect leading to wilting for seven days period at tillering, shooting and flowering stages on survival, tiller production dry matter accumulation in stem and leaf, ear length and number and yield components led to the conclusions that wilting at shooting and flowering caused progressive decrease in survival value and grain number, 1000 grain weight and grain yield. R.D. Asana and his coworkers reported the effect of intermittent soil drought during the entire life cycle and during grain development of wheat both under pot culture condition and field (Asana and Basu, 1963, Asana and Saini, 1958, Asana and Singh, 1967). Their results showed that under adequate soil moisture condition, ear number had the most potent effect on yield whereas under intermittent drought, grain number per ear and sometimes 1000 grain weight had as much effect as ear number. The average reduction due to water stress was 24% in ear number, 36% in grain number and only 9% in 1000 grain weight. The relatively greater vulnerability of grain number per year was apparently because water stress reduced spikelet as well as grain number (Asana and Mani, 1955,). It was further observed that earing capacity was not fully expressed under rainfed condition. The percentage reduction in ear number was 40 to 55, which was much larger than under pot culture.

In other experiments same workers have shown that early wheat varieties (NP 875 and NP 876) tillered less than other varieties which eared later by 2 to 4 weeks. The maximum tiller number per plant was however achieved in about 8 weeks from sowing in all varieties studied. Besides the main shoot, barely one tiller contributed to ear population in NP 823, NP 860 and NP 856 whereas in early varieties NP 875 and NP 876, the main shoot very largely contributed to the final ear population. Since the main shoot contributed very largely to grain yield under rainfed cultivation and genotype differences in grain number per ear and 1000 grain weight remained more or less constant, it appeared that selection for rainfed (barani) condition could be made on the basis of grain yield of the main shoot of the breeding material raised under irrigated conditions. Further studies have confirmed that ear number is the most important parameter for use in yield prediction when water supply is adequate (Hsu and Walton, 1970).

There are several reports on the selection criteria for the stress tolerance based on morphological, physiological and biochemical basis although the impact of these has been vary low. There are several likely reasons for lack of success. For example some criteria proposed have been related to more to survival mechanism under drought rather than productivity and some criteria are too difficult to measure and therefore are rarely adopted in any breeding programme. The limited success to date of the analytical approach may be due to lack of understanding of the physiological factors most directly involved in determining the yields, in addition to the absence of proper and easier method for evaluating them in a rapid and routine manner (Blum, 1988, Richard, 1996).

Laszlo *et al.* (2002) have applied simple and widely used field screening tests like water retention ability test of excised leaves and chemical dessication method. The water retention test indicates the cuticular resistance of the flag leaves (Clark, 1982) while by chemical dessication one can model a lethal post anthesis water stress and check the plants ability to support grain filling by translocated stem reserves. The method is still under test. One approach that has received more attention is the solute accumulation in plants or osmoprotection, which may confer water stress tolerance (Chen and Marata, 2002, McNeil *et al.* 1999). Several metabolites have been reported to be a part of osmoprotection such as proline, glycine betaine, mannitol, sorbitol, fructans, trehalose etc. and physiological research has already identified genetic differences in solute accumulation under water deficit condition in several crop species but the overall results of these studies have provided little success in improving the drought tolerance. Transgenic plants engineered to syntheze more of proline, mannitol, fructans, trehalose, glycine betaine exhibit marginal improvement in drought/salt tolerance.

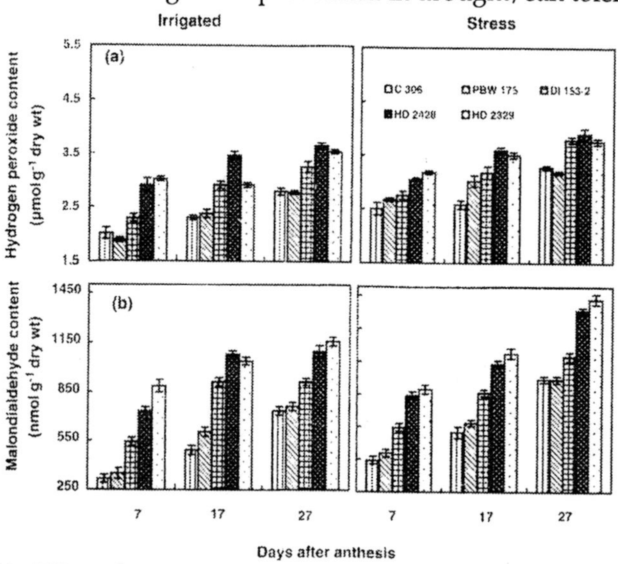

Fig. 10 : Effect of water stress on (*a*) hydrogen peroxide content and (*b*) lipid peroxidation (malondialdehyde content) in wheat genotypes Vertical bars show +_SE of mean; LSD significant at P=0.05
Source : Sairam and Srivastava (2001)

Stress particularly water stress induces oxidative stress which damages the plant cell (Fig 10). In wheat, the role of antioxidative enzymes such as superoxide dismutase, glutathione reductase, catalase has been shown to protect plants from oxidative stress (Fig. 11 and 12) (Sairam and Srivastava, 2001, Agarwal *et al.*, 2005) however, the protection is not very significant. Agarwal *et al.* (2005) have shown the role of salicylic acid (SA) induced H_2O_2 accumulation in germinating wheat seedling was not associated with inhibition of catalase or ascorbate peroxidase.

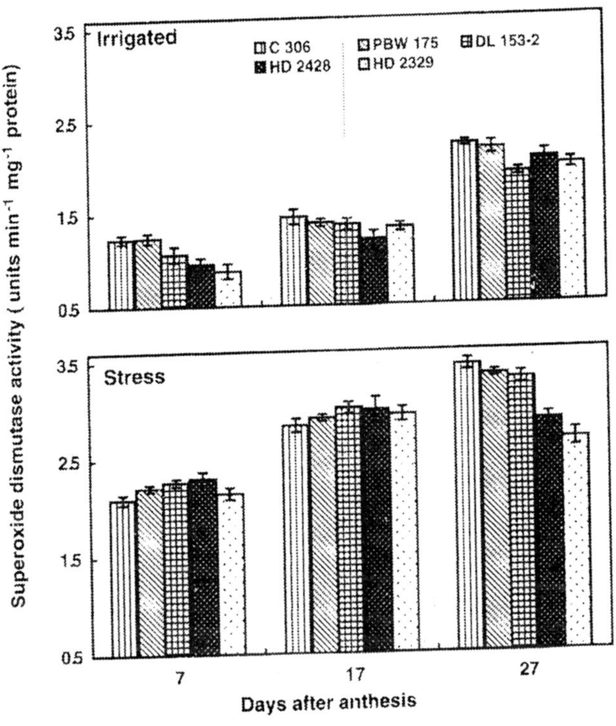

Fig. 11 : Effect of water stress on superoxide dismutase activity in wheat genotype. Vertical bars show ± S.E. of mean; LSD significant at P=0.05

Source : Sairam and Srivastava 2001.

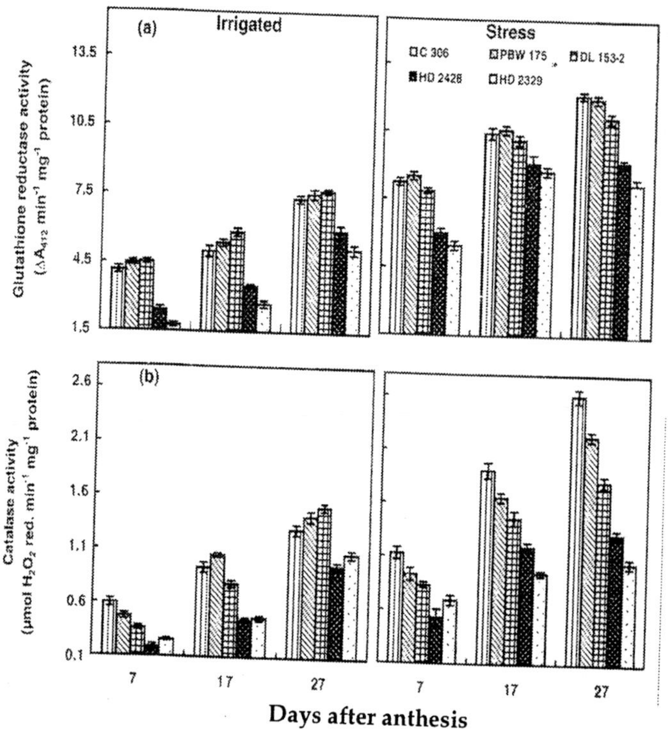

Fig. 12 : Effect of water stress on (*a*) glutathione reductase and (*b*) catalase in wheat genotype. Vertical bars show ± S.E. of mean; LSD significant at P=0.05

Source : Sairam and Srivastava, 2001.

They further suggested that abiotic stress signal is transduced via ABA, Ca^{2+} and H_2O_2 which might be responsible for the activation of some common transcription factor associated with SOD, APOX and CAT (Fig.13). There are many other characters morphological, physiological and biochemical which impart tolerance to plants and these are controlled by multiple genes. Thus, concentrating on one or two genes may not be of any significance in developing a drought resistant tolerant plant.

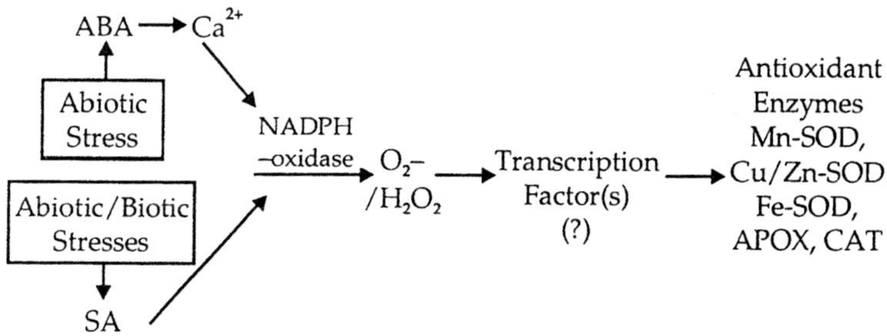

Fig. 13 : Sequence of events suggesting role of ABA, SA, Ca and H_2O_2 in the induction of antioxidant activity

Source : Agrawal *et al.*, 2005

Transcription factors have been shown to control the activity of multiple response genes in a coordinated manner and therefore, represent attractive targets for application in molecular plant breeding. Yamaguchi-Shinozaki and Shinozaki (2005) have suggested that several families of transcription factors were found to be involved in the regulation of drought response in plants. In wheat efforts have been made using biolistic bombardment. Transgene presence and expression in T_0 transgenic plants and or T_1 generation plants have been analyzed following southern and northern plot hybridization. The results after full analysis may possibly reveal the mechanisms of stress tolerance in wheat plants.

TEMPERATURE STRESS

Temperature is one of the important environmental factor affecting the growth and yield of crop plants. In wheat, the most important effect of temperature has been observed during ear development stage. Plants growing at lower temperature tend to have long culms, large flag leaf and potentially fertile florets in each spikelet (Warrington *et al.*, 1977). The grain growth duration from anthesis to maturity is significantly affected by rise in temperature. In one of the experiments, pooled data of thee wheat varieties, namely Kalyansona, HD 4502 and Sonalika it has been shown that between 13°C and 27°C, the grain growth duration from anthesis to maturity declined by 2.6 day^{-1} for every 1°C rise in temperature followed by a slow decline at higher temperature upto 30°C (Saini and Dadhwal, 1986a).

Temperature and photoperiod are two environmental variables, the impact of these two is used to calculate the changes in duration of flowering. The results showed that development was faster under warmer weather and longer photoperiod and the two variables interact to provide different levels of response in different genotypes. The impact coefficient of temperature, light intensity as mega joules m^{-2} day^{-1} of PAR) and duration of the phase from terminal spikelet to anthesis or grain number showed the increase by 226 grains/m^2 for every one day increase in duration above 40 days, 273 grain/m^2 for every IMJ/m^2 increase in PAR above its mean value of 8.7 and declined by 685 grain/m^2 for every 1°C resi in temperature above 17.7°C as orthogonal function during terminal spikelet formation to anthesis. Increase in light intensity increased grain number, its accompanying increase in temperature would tend to increase it.

While studying the mechanism of temperature effect in wheat, two contrasting varieties of wheat Yangmai 9 and Xuzhou 26, Zhao *et al.* (2008) reported that higher temperature. High temperature reduced both content and yield of starch in grains, while enhanced protein content and reduced protein yield in grains. High temperature significantly enhanced the activities of starch synthase (SS) and granule bound starch synthase (GBSS) on 14 days after anthesis (DDA). High temperature affected soluble starch synthase (SSS) slightly in Yangmai 9, but reduced SS activity markedly in Xuzhou 26 pm 14 DDA. However, at the middle and late stages of grain filling, high temperature reduced the activities of SS, GBSS and SSS significantly in he two wheat cultivars. The activities of SS, SSS and GBSS significantly correlated with starch accumulation in grains, except for GBSS activity to starch content on 14 DAA. Similar observation has been made by Sumesh *et al.* (2008) in two wheat varieties, a temperature sensitive WH 542 and a temperature tolerant HD 2285. The excised developing grains (20 d after anthesis) of ambient grown plants were incubated at 15, 25, 35 and 40°C for 2 hours and then analyzed for soluble starch synthase (SSS), kinetic parameters of SSS and content of heat shock protein (HSP 100). The elevated temperature during grain development significantly decreased grain growth in WH 542 whereas no such decrease was observed in HD 2285. High temperature tolerance of HD 2285 was found to be associated with high catalytic efficiency (V_{max}/K_{max}) of SSS at elevated temperature and higher content of HSP 100. HD 2285 had higher SSS activity and also lesser decrease in SSS activity under high temperature treatment compared to WH 542 (Table 7). SSS from HD 2285 had lower Km (ADPG) and higher Vmax than WH 542 under higher temperature treatment. HD 2285 maintained higher Vmax/Km at 25 and 40°C indicating greater efficiency of its SSS compared to SSS from WH 542. Activity of GBSS was considerably lower than SSS at different temperature probably because GBSS is bounded to granule. It has been suggested that for improving thermotolerance forr grain growth, there is need to incorporate thermostable form of SSS and overexpression of HSP 100 in a high yielding cultivar of wheat.

Table 7 : Soluble Starch Synthase (SS) and Granule Bound Starch Synthase (GBSS) Activity (nmol per gram dry matter) in the Excised Grains (20 DAA) Following Exposure to Different Temperatures in Wheat Cultivars. Mean ± SE n=3

Cultivars		*15° C*	*25°C*	*35°C*	*45°C*
HD 2285	SSS	299.54 ± 23.96	352.03 ± 10.08	243.34 ± 13.89	183.44 ± 15.22
WH 542		228.29 ± 13.62	243.95 ± 9.60	189.19 ± 13.89	92.64 ± 13.21
HD2285	GBSS	117.19 ± 4.51	109.26 ± 8.03	123.37 ± 7.10	95.62 ± 7.16
HD2285		121.29 ± 7.86	121.40 ± 4.62	104.35 ± 7.37	72.31 ± 9.73

Source : Sumesh *et al.*, 2008

The above observation that tolerance to high temperature for grain growth is associated with catalytic efficiency of soluble starch synthase suggests that this enzyme could be an important target for improving the high temperature tolerance for grain growth in wheat. The high sensitivity of this enzyme to high temperature has also been reported (Rijven 1986, Jenner 1994).

SALINITY STRESS

Amongst the abiotic stresses, salinity is one of the most significant and widespread stress of arid and semi arid lands. High concentration of salt have detrimental effects on seed germination, plant growth and yield. The salinity causes change in ionic content of the plant (Zaib-un-Nisa Abdullah *et al.*, 1978) resulting in manifestation of morphological and physiological attributes. The effect of salinity in wheat is not uniform throughout growth and development. The main effect of salinity was reported on floret abortion, and subsequent grain/spike and grain weight. Grain yield/spike got reduced by increasing salinity level and exposure duration (Kamkar *et al.*, 2004).

Variability for salt tolerance has been observed in wheat varieties. Most of the high yielding varieties can be grown in sodic soil only upto pH 9. Beyond this pH, only salt tolerant varieties have the ability to survive and produce some economic yield. For saline stress, wheat can be grown upto ECe 5.6 m mhos/cm.

In the last few years, a number of papers have appeared on the study of the salinity tolerance mechanism in plants particularly in wheat. Salinity stress exerts many symptoms similar to those observed under water deficit. Similar to water stress, reactive oxygen species (ROS) such as super oxide radical (O_2^-), hydrogen peroxide (H_2O_2) and hydroxyl radical ($OH^{\cdot}$) are also produced during salinity stress, and are responsible for the damage to membranes and other essential macromolecules such as photosynthetic pigments, proteins, DNA and lipids (Hermandez *et al.*, 1993, 1995, Fadzilla *et al.*, 1997). In plant cells chloroplasts, mitochondria and peroxisomes are important intracellular generators of ROS (Del Rio *et al.*, 1991, Elstner, 1991, Salin, 1991). ROS produced as a result of various abiotic stresses need to be scavenged for maintenance of normal growth. The primary scavenger is superioside dismutase (SOD), which converts O_2^- to H_2O_2 which is eliminated by ascorbate peroxidase (APOX) in association with dehydroascorbate reductase and glutathione reductase (GR) and regenerate the ascorbic acid (Salin 1991, Asada 1994). H_2O_2 is also scavenged by catalase (Dhindsa *et al.*, 1981, Anderson *et al.*, 1995, Comba *et al.*, 1998), though the enzyme is less efficient than ascorbate perioxidase-glutathione reductase system. Changes in activities of various antioxidant enzymes under salinity stress have been reported (*e.g.* Gueta-Dhan *et al.*, 1997, Fahmy *et al.*, 1998, Hernandez *et al.*, 2000).

Sairam and Srivastava (2002) studied the effects of long term, medium level (electrical conductivity of extract (ECe) = 6.85 dSm^{-1}) sodium chloride (NaCl) salinity to tolerant (Kharchia 65) and susceptible (HD 2687) wheat genotypes Salinity tolerant wheat cv. Kharchia 65 showed less decline in RWC, CHL, MSI estimated in whole tissue than salt sensitive HD 2687. Kharchia 65 also exhibited less decrease in AA content, less increase in H_2O_2, TBARS contents (Table 8.) and higher increase in SOD and its isozymes, APOX and GR in all sub-cellular fractions than salt sensitive HD 2687. H_2O_2, TBARS contents and AA contents were higher in chloroplastic fraction. Chloroplastic fraction showed higher total SOD, APOX and GR activity, followed by mitochondrial fraction in case of total SOD and GR, while cytosolic fraction was in second place in case of APOX activity (Fig. 14 and 15).

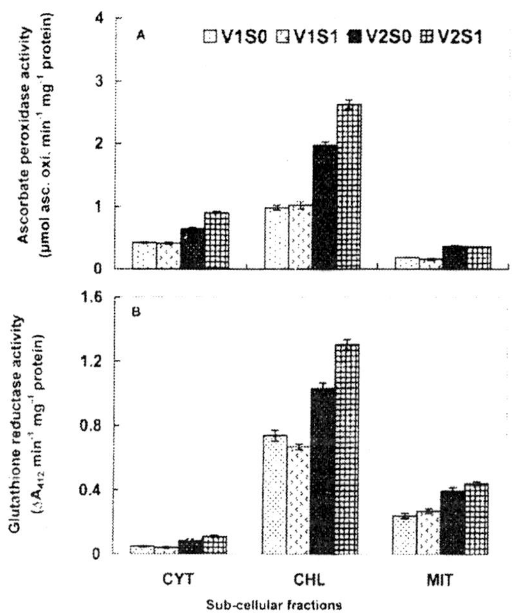

Fig. 14 : Effect of salinity stress on APOX (A) and GR (B) activity in different cell factions of tolerant and susceptible wheat genotypes. Vertical bars indicate ±S.E. of mean. Data significant at P=0.05 (n=16). V1, HD 2687; V2, Kharchia 65; S0, control (ECe = 1.7 dS m^{-1}), S1, 100 mM NaCl (ECe = 6.85 dS m^{-1}).

Source : Sairam and Srivastava (2002).

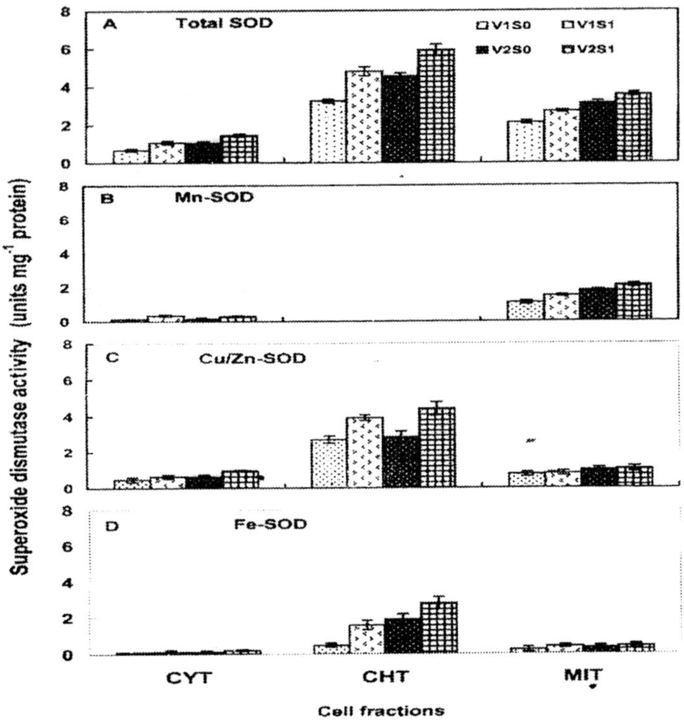

Fig. 15 : Effect of salinity stress on the activity of total SOD (A), Mn-SOD (B), Cu/Zn-SOD (C) and Fe-SOD (D) in different cell factions of tolerant and susceptible wheat genotypes. Vertical bars indicate ±S.E. of mean. Data significant at P=0.05 (n=16). V1, HD 2687; V2, Kharchia 65; S0, control (ECe = 1.7 dS m^{-1}), S1, 100 mM NaCl (ECe = 6.85 dS m^{-1}).

Source : Sairam and Srivastava (2002)

Sairam *et al.* (2005) have further reported that constitutive as well as salt induced increase in antioxidative enzyme are important in providing protection against ROS. While constitutive level provide protection from oxidative stress arising form normal oxidative metabolism, the salinity or abiotic stress induced increase in antioxidative enzyme in response to increase in oxidative stress actually decide the level of tolerance in plants.

Breeding for improved salt tolerance and other remedial measures such as reclamation of soil may help in improving crop yield under saline condition.

Table 8. Effect of salinity on H_2O_2, lipid peroxidation (TBARS) and AA Content in Tolerant (Kharchia 65) and Sensitive (HD 2687) Wheat

Genotype	*Hydrogen Peroxide Content (μmol g⁻¹ dry wt.)*		*Thiobarbituric acid reactive substances (μmol g⁻¹ dry wt.)*		*Ascorbic acid (μmol g⁻¹ dry wt.)*	
	Cont.	*Salinity*	*Cont.*	*Salinity*	*Cont.*	*Salinity*
HD 2687	2.803(±0.123)	3.042(±0.107)	972.08(±21.96)	1097.03(±22.16)	34.88(±1.48)	20.97(±1.07)
Kharchia 65	1.807(±0.104)	2.363(±0.064)	598.87(±28.78)	724.90(±20.50)	53.15(±2.03)	34.67(±0.98)

Figures in parenthesis denotes ±S.E. of mean, (n=16).

Source : Sairam and Srivastava (2002)

SCREENING FOR TOLERANCE

Many potential trait or criteria have been proposed for screening salt tolerance. In many cases the test criteria is an unrealistic predictor of field performance. For example, germination test done on filter paper in saline condition may not give a true picture in field experiment. A number of workers have suggested that screening for salt tolerance be carried out using physiological markers or that physiological traits should be used as selection criteria either singly or in combination (pyramiding) rather than selection for yield or yield components (Shannon and Noble, 1990, Flowers and Yeo, 1995). Plants limiting the uptake of toxic ions or maintaining normal ion contents such as Na^+ and K^+ could be useful selection trait. Selection within varieties of lines with low Na^+ transport has been accomplished in rice (Yeo *et al.*, 1988). In wheat salt tolerance is associated with enhanced ability to discriminate between Na^+ and K^+ in the soil solution and to preferentially accumulate K^+ and exclude Na^+ (Gorham *et al.* 1985, Omielan and Epstein 1991). Durum wheat lacks D genome of bread wheat; tend to accumulate more Na and less K^+ than bread wheat under salinity stress. The trait for K^+/Na^+ discrimination affects transport of K^+ and Na^+ to the shoot with little effect on root ion concentration or anion concentration in the leaves and the main site of action is thought to be at xylem loading in the roots. It acts at all salt concentrations.

Other physiological mechanism such as ion accumulation (Flowers *et al.*, 1977). Osmotic adjustment (Morgan, 1977, Shannon 1987) and production of organic solutes (Greenway and Munns, 1980) have not yet shown the potential to be used as selection tools for salt tolerance in wheat. Other traits while not contributing in themselves to tolerance have been proposed for screening. For example, the induction and quenching kinetics of chlorophyll 'a' have been assessed. Krishnaraj *et al.* (1993) compared 3 week old wheat seedling of *Kharchia* and the salt sensitive variety *Fielder* at a range of salinity and found that the maximum rate of fluorescence induction and quenching decreased much more in *Fielder* than in *Kharchia* as salinity increased, and could be used in breeding programme to assess the salinity tolerance of wheat genotypes.

PLANT IDEOTYPE CONCEPT

Ideotype means an idealized envisioned appearance that is desired. Donald (1968) proposed the ideotype approach to plant breeding in contrast to the empirical breeding approach of defect elimination and selection for yield *per se*. He defended "crop ideotype" as an idealized plant with a specific combination of characteristics favourable for photosynthesis, growth and grain production based on knowledge of plant and crop physiology and morphology. He argued that it would be more efficient to define a plant type that was theoretically efficient and then breed for this.

There has been several reports to design and develop a new plant type in wheat. After several field experiments, Asana (1968) suggested a wheat ideotype for barani (rainfed) condition with the following main features.

Root

Must be substantial in 3rd and 4th foot of soil layer and probably related to time of earing.

Leaves

About seven on main axis (with a large flag leaf) preferably horizontally disposed.

Peduncle

The length of ear bearing internode should form a substantial proportion of the length as photosynthesis in peduncle and flag leaf sheath contribute significantly to yield.

Ear

Must have fertile spikelets or preferably branched ear so that a large number of grain can develop.

Time of Earing

Ear must appear at such a time that grain can develop for at least five weeks at a mean temperature of 25°C.

Asana's group working on wheat physiology observed that the tillering potential of wheat was not fully expressed under rainfed condition. Hence the main axis produced most of the ear population. The main axis yield (grain number and weight) therefore had an important influence on yield. The finding is of practical significance as it implies that selection of varieties for non irrigated condition be based on yield of main axis ear from a population from irrigated condition (Asana, 1969).

For achieving quantum jump in wheat productivity under optimum water and nutrient condition, CIMMYT has suggested Agropolitetra wheats which will have intermediate tillering capacity (upto 10 tillers), long spike (30 cm), high fertility (upto 200 grains) broad leaves and thick stems as their basic visual morphology. Most of these characters have already been combined, except good and plump grains, especially in the short growing period of the Yagui Valley Sonora, Mexico. (Rajaram and Reynolds, 2001).

In an effort to achieve the quantum jump in wheat productivity, Singh (2001) has advocated to design a new plant type combining negatively correlated yield component in a single genotype of a very high yield potential. In this direction, a new plant type combining two negatively correlated traits (high tiller number with high grain weight and high grain weight with high grain number per ear) has been successfully developed resulting in significantly higher yield. The newly designed plant type has characteristic features of moderate tillering (375 per m^2 and above) all productive tillers, 50 and above grains per spike, grain weight between 45-50 g per 1000 grain, higher biomass, dark green, thick and broad leaves, thick stem, maturity duration between 120 to 135 and plant height between 85 and 100 cm and good root system. Efforts are on to increase the productive tillers with diverse genes for resistance to diseases in order to break the yield barrier (Singh *et al.*, 2005). The result of these developments has yet to come in the farmers field.

REFERENCES

Agarwal, S., Sairam, R.K.,'Srivastava, G.C., Tyagi, Aruna and Meena, R.C. (2005). Role of ABA, salicylic acid, calcium and hydrogen peroxide on antioxidant enzymes induction in wheat seedlings. *Plant Sci.*, **169**: 559-570.

Akiko, Kozaki and Takeba G.O. (1996). Photorespiration protects C_3 plants from photooxidation.Nature 384,557-560.

Anderson, M.D., Prasad, T.K., Stewart, C.R. (1995). Changes in isozyme profiles of catalyse, peroxidase and glutathione reductase during acclimation to chilling in mesocotyls of maize seedlings. *Plant Physiol.*, **109**: 1247-1257.

Andrews, M. Lea, P.J Raven, J.A and Lindsey, K (2004). Can genetic manipulation of plant assimilation enzymes results inm increase crop yield and geater nitrogen use efficiency. An assessment. Ann Apl. Biol. 145 (1) 25-40.

Asada, K. (1994). Production and action of active oxygen in photosynthetic tissues. In : Foyer, C.H., Mullineaux, P.M. (ed.) : Causes of Photooxidative Stress in Plants and Amelioration of Defence System. Pp. 77-109. CRC Press, Boca Raton, 1994.

Asana, R.D (1961). Analysis of drought resistance in wheat in plant water relationship in arid and semi arid conditions. Arid zone research 16, 183-190.UNESCO, Paris.

Asana,R.D (1969).Plant type for drought condition. J. Post Graduate School 7 : 38-39. IARI, New Delhi

Asana, R.D. and Basu, R.N. (1963). Studies in physiological analysis of yield. VI-Analysis of the effect of water stress on grain development in wheat. *Indian J. Pl. Physiol.*, **6** : 1-13.

Asana, R.D. and Mani, V.S. (1955). Studies on physiological analysis of yield I. Varietal differences in photosynthesis in leaf stem, ear and wheat. *Physiol. Plant*, **3**: 22-39.

Asana, R.D. and Saini, A.D. (1958). Studies in physiological analysis of yield. IV-The influence of soil drought on grain development, photosynthetic surface and water content of wheat. *Physiol. Plant*, **11** : 666-674.

Asana, R.D. and Singh, D.N. (1967). On the relationship between flowering time root growth and soil moisture extraction in wheat under non irrigated cultivation. *Indian J. Pl. Physiol.*, **10** : 154-169.

Austin, R.B. (1990). Prospects for genetically increasing the photosynthetic capacity of crops. In perspectives in biochemical and genetic regulation of photosynthesis. P305-409 NY, USA, Allan R. Liss.

Austin, R.B., Morgan, C.L., Ford, M.A. and Bhaghat, S.G. (1982). Flag leaf photosynthesis of *Triticum aestivum* and related diploid and tetraploid species. *Ann. Bot.,* **49** : 177-189.

Baker,C.K and Gallagher J, N. (1983 b). The development of winter wheat in the field.2.Control of primordium initiation rate by temperature and photoperiod. J. Agric. Sci. Camb 101 : 337-344..

Beevers, L. and Hageman, R.H. (1969). Nitrate reduction in higher plants. *Ann. Rev. Pl. Physiol.,* **20**: 495-522.

Bernier, G, Havelange,A, Housa, C, Petitjean, A and Lejeune, P (1993). Physiological signals that induce flowering.Plant Cell 5,1147-1155.

Bird, I.F., Cornelius, M.J. and Keys, A.J. (1982). Affinity of RuBP carboxylase for carbon dioxide and inhibition of the enzyme by oxygen. *J. Exp. Bot.,* **33**: 1004-1013.

Blackman V.H. (1919). The compound interest law and pant growth. Ann. Bot. 33 : 353-360.

Blum, A. (1988). Plant breeding for stress environments. CRC Press, Florida, USA.

Cai, C., Zhao, X.Q., Zhu, Y.G., Li, B., Tong, Y.P. and Li, Z.S. (2007). Regulation of the high-affinity nitrate transport system in wheat roots by exogenous abscisic acid and glutamine. *Journal of Integrative Plant Biology,* **49**(12) : 1719-1725.

Chandler, R.F. Jr. (1972). The impact of improved tropical plant type on rice yield in south and south-east Asia. pp. 77-85. In: rice breeding Intl. Rice Instt. Los Banos Philippines.

Chen T.H.H and Murata N (2002) Enhancement of tolerance of abiotic stress by metabolic engeneering of betains and othercompatable solutes.Current Openion in Plant Biology 5(3)250-257.

Chinoy, J.J. (1942). Pre-sowing treatment and phasaic development. *Curr. Sci.,* **11** : 400.

Chinoy, J.J. (1960). Physiology of drought resistance in wheat. I-Effect of wilting at different stages of growth on survival values of eight varieties of wheat belonging to seven species. *Phyton.,* **VI** : 147-157.

Chinoy, J.J. (1962). Formation and utilization of ascorbic acid in the shoot apex of wheat as factors of growth and development. Indian J.Pl. Physiol. 5 : 172-201.

Chinoy, J.J., Abraham, P.G., Pandya, R.B., Saxena, O.P., Dave, I.C. (1970). Effect of presowing treatment of titicum seeds with ascorbic acid and sucrose on growth, development and yield characters. *Indian J. Pl. Physiol.*, **13** : 40-48.

Clarke, J. (1982). Excised leaf water retention capacity as an indicator of drought resistance of triticum genotypes. *Can. J. Plant Sci.*, **62**: 571-576.

Comba, M.E., Benavides, M.P., Tomaro, M.L. (1998). Effect of salt stress on antioxidant defence system in soybean root nodules. *Aust. J. Plant Physiol.*, **25**: 665-671

Crafts-Brandner, S.J. and Salvucci, M.E. (2000). Rubisco activase constrains the photosynthetic potential of leaves at higher temperature and CO2.Proc.Natl.Acad.Sci.USA 21 97(24) 13430-13435.

Craver, B.F and Nevo, E (1990). Genetic diversity of photosynthesis characteristics in native population of Triticum dicoccoides. Photosynthesis res. 25(2) 119-128.

Daniel-Vedele, F., Filleur, S., Caboche, M. (1998). Nitrate transport: a key stem in nitrate assimilation. *Current Opinion in Plant Biology*, **1** : 235-239.

Decker, J.P. (1957). Further evidence of increased carbon dioxide production accompanying photosynthesis. J Solar Energy Sci Eng, 1 : 30-33.

Decker, J.P. and Tio, M.A. (1959). Photosynthetic surges in coffee seedlings. In Journal of Agriculture of the University of Puerto Rico. 43 : 50-55.

Del Rio, L.A., Sevilla, F., Sandalio, L.M., Palma, J.M.L. (1991). Nutritional effects and expression of superoxide dismutase: induction and gene expression, diagnostics, prospective protection against oxygen toxicity. *Free Radical Res. Commun.*, **12-13**: 819-828.

Dhanda, S.S., Sethi, G.S. and Behl, R.K. (2004). Induces for drought tolerance in wheat genotypes at early stages of plant growth. *J. Agr. Crop Sci.*, **190** : 6-12.

Dhindsa, R.A., Plumb-Dhindsa, P., Thorpe, T.A. (1981). Leaf senescence: correlated with increased permeability and lipid peroxidation, and decreased levels of superoxide dismutase and catalase. *J. Exp. Bot.*, **126** : 93-101. Donalde,C.M. (1968). The breeding of crop ideotypes. *Euphytica*, **17** : 385-403.

Donalde,C.M. (1968). The breeding of crop ideotypes. *Euphytica*, **17** : 385-403.

Duston, R.L., Gifford, M. and Evaus, L.T. (1973). Photosynthesis characteristics of modern and primitive wheat species in relation to ontogeny and adoption to light. *Aust. J. Biol. Sci.*, **26** : 295-307.

Ehdaie, B. (1995). Variation in water use efficiency and its components in wheat. II. Pot and field experiments. *Crop Sci.*, **35** : 1617-1626.

Elstner, E.F. (1991). Mechanism of oxygen activation in different compartments of plant cell. In: Pell, E.J., Steffen, K.L. (ed.) Active Oxygen/Oxidative Stress and Plant Metabolism. Pp. 13-25. *Amer. Soc. Plant Physiol.*, Rockville, 1991.

Evans,L.T and Rowson,H.M (1970). Photosynthesis and respiration in flag leaf and component of ear during grain development in wheat. Aust. J. Biol. Sci. 23, 245-254.

Fadzilla, N.M., Robert, P., Finch, R.P., Burdon, R.H. (1997). Salinity, oxidative stress and antioxidant response in shoot cultures of rice. *J. Exp. Bot.*, **48** : 325-331.

Fahmy, A.S., Mohamed, T.M., Mohamed, S.A. and Saker, M.M. (1998). Effect of salt stress on antioxidant activities in cell suspension cultures of cantaloupe (*Cucumis melo*). *Egyptian J. Physiol. Sci.*, **22** : 315-326.

Filowers, T.J., Troke, P.F. and Yeo, A.R. (1977). The mechanism of slat tolerance in halophytes. *Ann. Rev. Pl. Physiol.*, **28** : 89-121.

Flowers, T.J. and Yeo, A.R. (1995). Breeding for salinity resistance in crop plants : Where Next? *Aust. Plant Physiol.*, **22** : 875-884.

Friend, D.J.C., Helson, V.A. and Fusher, J.E. (1962). Leaf growth in morgues wheat as regulated by temperature light intensities and dry length. *Can. J. Bot.*, **40** : 299-311.

Gallagher, J.N. and Biscoe, P.V. (1978). A physiological analysis of cereal yield. II. Partitioning of dry matter. *Agric. Prog.*, **53** : 51-71.

Ghildiyal, M.C. (1988) Alteration of Ribulose bisphosphate carboxylase/ oxygenase ratio by urea. Curr. Sci. 57 : 200-201.

Ghildiyal, M.C. and Sirohi, G.S. (1986). Photosynthesis and sink efficiency of different species of wheat. *Photosynthetica*, **20**: 102-106.

Gorham, McDonnell, E., Budrewics, E. and Wyn Jones, R.G. (1985). Salt tolerance in the triticease. Growth and solute accumulation in leaves of *Thinopyrum hessarabicum*. *J. Expt. Bot.*, **36**: 102-103.

Greenway, H. and Munns, R. (1980). Mechanism of salt tolerance in non halophytes. *Ann. Rev. Pl. Physiol.*, **31**: 149-190.

Gueta-Dahan, Y.., Yaniv, Z., Zilinkas, B.A., Ben-Hayyim, G. (1997). Salt and oxidative stress: similar and specific responses and their relation to salt tolerance in citrus. *Planta*, **203**: 460-469.

Hay, R.K.M. and Kirby, E.J.M. (1991). Convergence and synchrony a review of the coordination of development in wheat. *Aust. J. Agric. Res.*, **42**: 661-700.

Hernandez, J.A., Corpas, F.J., Gomez, M., Del Rio, L.A., Sevilla, F. (1993). Salt induced oxidative stress mediated by activated oxygen species in pea leaf mitochondria. *Physiol. Plant,* **89** : 103-110.

Hernandez, J.A., Jimenez, A., Mullineaux, P. and Sevilla, F. (2000). Tolerance of pea (*Pisum sativum* L.) to long term salt stress is associated with induction of antioxidant defences. *Plant Cell Environ.,* **23** : 853-862.

Hernandez, J.A., Olmos, E., Corpas, F.J., Sevilla, F., Del Rio, L.A. (1995). Salt induced oxidative stress in chloroplast of pea plants. *Plant Sci.,* **105** : 151-167.

Holbrook,G.P.; Keys, A.J and Leech, R.M (1984). Biochemistry of photosynthesis in species of triticum of different ploidy. Plant Physiol. 74, 12-15.

Hsu, P. and Walton, P.D. (1970). The inheritance of morphological and agronomic characters in spring wheat. *Euphytica,* **19** : 54-60.

Jain, H.K. (1986). Eight years of post mendalian breeding for crop yield nature of selection presumes and future potential. *Indian J. Genet.,* **46**: (Supplement) 30-53.

Jenner, C.F. (1994). Starch synthesis in kernel of wheat under high tremperature condition. Aust. J. Pl. Physiol. 21 : 791-806.

Jones, R.J.; Roussler, J and Quattar, S (1985). Thermal environment during cell division in maize:effects on number of endosperm cells and starch granules. Crop Sci. 25, 830-834.

Jordan, D.B. and Ogren, W.l (1981). A sensitive assay procedure for simultaneous determination of Ribulose 1-5 bisphosphate Carboxylase and Oxygenase activities. Plant Physiol 67(2). 237-245.

Kamkar, B., Kafi, M. and Mahallati, A.N. (2004). Determination of most sensitive development period of wheat (*Triticum aestivum*) to salt stress to optimize saline water utilization. 4th International Crop Science Congress, Iran.

Karim, M.M. and Siddique, M.M. (1991). Crop growth and relative growth rate of old and modern wheat cultivars. *Austr. J. Agric. Res.,* **42** : 13-22.

Keys, A.J.; Sampio, Comolius, M.J. and Bird, I.F. (1977). Effect of temperature on photosynthesis and photorespiration of wheat leaves. J. Exp. Bot 28 : 525-533.

Keeling, P.L.Banisadr, R; Barone, L; Vasserman, B.P.; Singletary, G.W. (1994). Effect of temperature on enzymes in the pathway of starch biosynthesis in developing wheat and maize grain. Aust. J.Pl. Physiol. 21 : 807-827.

King, R.W.; Wardlaw, I.F. and Evans, L.T. (1967). Effect of assimilate utilization on photosynthetic rate in wheat. Planta. Berl. 77, 261-267.

Kirby, E.J.M. and Appleyard, M. (1984). Cereal development guide. Stonelergh Kenilworth, UK, NAC Arabable Unit pp. 95.

Krishnaraj, Mawson, B.T.; Yeung, E.C. and Thorpe, T.A (1993). Utilisationofinduction and quenching kinetics of chlorophyll a fluorescence for *in vivo* salinity screening studies in wheat (*Triticum aestivum* var Kharcha 65 and Fielder). Can J. Bot 71 : 87-92.

Kulshreshtha, V.P. and Jain H.K. (1982). Eighty years of wheat breeding in India,past selection pressure and future prospects. Zeitschrift fur Pflanzenzuchtung. 89, 19-30.

Laszlo Cseuz, Janos Pauk, Zoltan Kertesz, Janos Matuz, Peter Fonad, Irma Tari and Laszlo Erdei (2002). Wheat breeding for tolerance to drought stress at the cereal research non-profit company. *Acta Biologica Szegediensis,* **46**(3-4) : 25-26.

Lommersville, C.R. (1986). Future prospects for genetic manipulated of RUBISCO. *Phil. Trans. Royal Sco. Lond.,* **313** : 459-468.

Long, S.P. (2006). Low growth temperature modify the efficiency of light use by photosyntem II for chilling tolerant C_4 species. *Cyperus longus* L. and *Miscantheus giganteus. Plant Cell Environ.,* **29** : 720-728.

Loomis, R.S. and Amthor, J.S. (1996). Limit of yield revisited. In : M.P. Reynolds, S. Rajaram and A. McNab (Eds.). Increasing yield potential in wheat breaking the barriers. P76-89, Mexico D.F. CIMMYT.

Major, D.J. and Kiniry, J.R. (1991). Predicting daylight effects on phonological process. In : T. Hodges (ed.) Predicting Crop Phenology pp. 15-28. Boca Raton, Fl. USA CRC Press.

McCree, K.J. (1974). Equation for the rate of dark respiration of white clover and grain sorghum as function of dry weight, photosynthetic rate and temperature. *Crop Sci.,* **14** : 509-514.

McNeil, S.D.; Nucci, M.L. and Hanson A.D. (1999). Betains and related osmoprotectants. Targets for metabolic engeneering of stress resistance. Pl. Physiol. 120 : 945-949.

Mian, A.R. and Nafziger, E.D. (1994). Seed size effects on emergence, head number and grain yield of winter wheat. *J. Prod. Agric.,* **5** : 265-268.

Monteith, J.L. (1977). Climate and the efficiency of crop production in Britain Philosophical Transaction of the Royal Society of London, **281** : 277-294.

Morgan, J.M. (1977). Differences in osmoregulation beween whaet cultivars. Nature (London) 270, 234-235.

Morgan, J.A. and Lecain,D.R (1991). Leaf gas exchange and related leaf traits among 15 winter wheat genotypes. Crop Sci 31 : 443-448.

Moss, D.N. (1966). Respiration of leaves in light and darkness. Crop Sci 6 : 351-354

Natr, L. (1966) Varietal difference in the intensity of photosynthesis. (Russian) Rostl. vyroba 12, 163-178.

Neiles, T.F. and Incoll, L.D. (1968). The control of leaf photosynthesis rate by the level of assimilate concentration in the leaf. A review of the hypothesis. *Bot. Rev.*, **34** : 107-125.

Ogren, W.L. (1994) Photosynthesis : pathways,regulation and modification. Ann. Rev. Plant Physiol. 35, 514-442.

Omielan, J.A. and Epstein, E. (1991). Salt tolerance and ionic relations of wheat as affected by chromosomes of salt tolerant *Lophopyrum elongatum. Genome.*, **34** : 961-974.

Plorgan, J.M. (1977). Differences in osmoregulation between wheat cultivars. *Nature, London,* **270** : 234-235.

Planchon, C (1969). Photosynthetic activity and yield in soft wheat (Triticum aetivum). Genet.agri 23 : 485-490.

Rachmileviteh, S., Cousin, A.B. and Bloom, A.J. (2004). Nitrate assimilation in plant shoot depends on photorespiration. *Plant Biol.,* **101**(31) : 11506-11510.

Radosevich, S.R. and Osteryoung, K. (1987). Principles governing plant environment interactions. In: Forest vegetation management for confer production. Wasktad, J.D. and P.J. Kuch (Eds.). John Wiley and Sons. New York, pp. 523.

Rahman, M.S., Wilson, J.H. and Aitken, V. (1977). Determination of spikelet number in wheat II. Effect of varying light level on ear development. *Aust. J. Agric. Res.,* **26** : 575-581.

Rajaram, S. and Reynolds, M.P. (2001). International wheat breeding. Plant Breeding: Sustaining the future. Abstracts of the XVI Eucarpia Congress Edinburgh Scotland Sept. 10-14.

Rawson, H.M. and Evans, L.T. (1971). The contribution of stem reserves to grain development in a range of wheat cultivars of different height. *Aust. J. Agric. Res.,* **22** : 851-863.

Read, B.A. and Tabita, F (1994). Higher substrate specificity factor ribulose bisphosphate carboxylase/oxygenase from eucaryoticmarine 'algae' and properties of recombinant cynobacterial Rubisco containingalgal residue modifications.Arch. Bichem Biophys 312 : 210-218.

Reuveni, J. and Bighee, B. (1997) Very high CO2 reduces photosynthesis, dark respiration and yield in wheat. Ann. Bot. 80(4), 539-546.

Richards, R.A. (1996). Defining selection criteria to improve yield under drought. *Plant Growth Reg.*, **20** : 157-166.

Rijven, A.H.G.C. (1986). Heat activation of starch synthase in wheat endosperm tissue. Plant Physiol. 81 : 448-453.

Robson, M.J. (1982). The growth and carbon economy of a selection line of folium perenne cv. 523 with differing rates of dark respiration. *Ann. Bot.*, **49** : 321-329.

Saini, A.D. and Dadhwal, V.K. (1986a). Influence of sowing date on grain growth duration and karnel size in wheat. *Indian J. Agri. Sci.*, **45** : 210-212.

Saini, H.S. and Aspinall, D. (1982). Abnormal sporogenesis in wheat (*Triticum aestivum* L.) induced by short period of high temperature. *Ann. Bot.*, **49** : 835-846.

Sairam, R.K. and Srivastava, G.C. (2001). Water stress tolerance of wheat (*Triticum aestivum* L.) variation in hydrogen peroxide accumulation and antioxidant activity in tolerant and susceptible genotypes. *J. Agron. Crop Sci.*, **186** : 63-70.

Sairam, R.K. and Srivastava, G.C. (2002). Changes in antioxidant activity in sub alular fraction of tolerant and susceptible wheat genotypes in response to long term salt stress. *Plant Sci.*, **162** : 897-904.

Sairam, R.K., Srivastava, G.C., Agarwal, S. and Meena, R.C. (2005). Differences in antioxidant activity in response to salinity stress in tolerant and susceptible wheat genotypes. *Biol. Plant*, **49**(1) : 85-91.

Salvucci, M.E., Portis, A.R. Jr., Orgren, W.L. (1985). A soluble chloroplast protein catalyzes ribuloxbisphosphate carboxylase/oxygenase activation *in vivo*. *Photosynthesis Res.*, **7** : 193-201

Salin, M.L. (1991). Chloroplast and mechanism for protection against oxygen toxicity. *Free Radical Res. Commun.*, **12-13** : 851-858.

Salunke, M.R. and Asana, R.D. (1971). Comparative study of the development of grain in normal and branched ear types of wheat (*Triticum aestivum* L.). *Indian J. Agri. Sci.*, **41** : 1050-1054.

Samish, Y.B Pallas, J.E. Jr, Dornhoff, G.M. and Shibles, R.M. (1972). A re-evaluation of soybean leaf photorespiration. Plant Physiol 50 : 28-30.

Salvucci, M.E., Portis, A.R. Jr., Orgren, W.L. (1985). A soluble chloroplast protein catalyzes ribuloxbisphosphate carboxylase/oxygenase activation *in vivo*. *Photosynthesis Res.*, **7** : 193-201.

Sawada, S. (1970). An ecophysiological analysis of the difference between the growth rate of young wheat seedlings grown in various seasons. *J. Fac. Sci. Univ. Tokyo Sec. III Bot.*, **10** : 233-263.

Shannon, M.C. (1987). Salinity-An environmental constraints on crop productivity. In : Agronomy 1987-Responding to change. Proc. 4th Australian Agronomy Conference 9-18 La Trobe Univ. Melbourne, Victoria, Aug.

Shannon, M.C. and Noble, C.L. (1990). Genetic approaches for developing economic salt tolerant crops. Pp. 161-185. In : K.K. Tanji (ed.) Agricultural salinity assessment and management. Am. Soc. Civil Eng. New York.

Shiroya, M. (1977). Translocation of organic substances in sunflower I. Downward translocation of ^{14}C same. *Plant Cell Physiol.*, **18** : 625-639.

Singh, S.S. (2001). Breaking yield, barring in wheat-new plant type designed. *Wheat Inf. Serv.*, **93** : 22-26.

Singh, S.S., Sharma, J.B., Sharma, D.N. and Chand, N. (2005). Wheat production in stressed environments. Proc. 7th International Wheat Conference 27 Nov.-2 Dec. Mardel Plata, Argentina.

Singh, V.P. and Arora, A. (2001). Intraspecific variation in nitrogen uptake and nitrogen utilization efficiency in wheat (*Triticum aestivum* L.). *J. Agron. Crop Sci.*, **186** : 239-244.

Somerville, C.R. (1986) Analysis of photosynthesis with mutants of higher plants and algae. Ann. Rev. Pl. Physiol. 37, 467-507.

Stoy,V (1965). Photosynthesis, respiration and carbohydrate accumulation in spring wheat in relatin to yield, Physiol, Plantar. Suppl, 16 : 178-182

Sumesh, K.V. Sharma-Natu, P and Ghildiyal, M.C. (2008). Starch synthase activity and heat shock protein in relation to thermal tolerance of developing wheat grains. Biol. Plant. 52(4) 749-753.

Sunmons, S.R. and Crookston, R.K. (1979). Rat and duration of growth of kernels formed at specific florets in spikelets of phring wheat. *Crop Sci.*, **19** : 690-693.

Swada, S (1970). An ecophysiological analysis of the difference between the growth rates of young wheat seedlings grown in various seasons. J. Fac. Sci, Univ. Tokyo Sect III Bot. 10, 233-263.

Talling J.F. (1961). Photosynthesis under natural condition. Ann. Rev Plant Physiol. 12, 133.

Todd, G.W. (1982). Photosynthesis and respiration of vegetative and reproductive parts of wheat and barley plants in response to increasing temperature. *Proc. Okla. Acad. Sci.*, **62** : 57-62.

Uemura, Anwaruzzaman Miyachi, S. and Yakota, A. (1997). Ribulose-1-5-bisphosphate carboxylase/oxygenase from thermophillic red algae with astrong specificity for CO_2 fixation. *Biochem. Biophysic. Res. Commun.* **233** : 568-571.

Wada, M., Luiz, J.C.B. Carvalho, Gustavo C. Rodrigues and Ryuichi Ishii (1994). Cultivar differences in leaf photosynthesis and grain yield of what under soil water deficit conditions. *Jpn. J. Crop Sci.,* **63**(2) : 339-344.

Warrington, F.J., Dunstone, R.L. and Sreen, L.M. (1977). Temperature effects at three development stages on the yield of the wheat ear. *Austr. J. Agric. Res.,* **28** : 11-27.

Watson, D.J. (1952).The physiological basis of variation in yield.Adv Agrnom 4 : 101-145

Woodend, J.J., Glan, D.M. and Person, C.U. (1986). Intraspecific variation for nitrate uptake and nitrogen utilization in wheat (*T. aestivum* L.) grown under nitrogen stress. *J. Plant Nutr.,* **9** : 1213-1225.

Yamaguchi-Shinozaki, K. and Shinozaki, K. (2005). Organization of cis-acting regulatory elements in osmotic- and cold-stress-responsive promoters. *Trend in Plant Science,* **10** : 88-94.

Yeo, A.R., Yeo, M.E. and Flavers, T.J. (1988). Selection of lines with high and low sodium transport from within varieties of an inbreeding species, rice (*Oryza sativa* L.). *New Physiologist,* **110**: 13-19.

Zaffaroni, E. and Schneiter, A.A. (1989). Water-use efficiency and light interception of semidwarf and standard-height sunflower hybrids grown in different row arrangements. *Agron. J.,* **81** : 831-836.

Zaib-un-Nisa,A Ahmed,R and Ahmed,J (1978).Salinity induced changes in the reproductive physiology of wheat plants. Plant and Cell Physiol. 19(1) 99-106.

Zelitch, I (1971) Photosynthess, Photorespiration and Plant Productivity. Academic Press, New York, pp. 140-169.

Zelitch, I and Day, P.R. (1968). Variation in photorespiration. The effect of genetic differences in photorespiration on net photosynthesis in tobacco. Plant Physiol. 43 : 1838-1844.

Zhao, H., Dai, T., Jiang, D. and Cao, W. (2008). Effects of high temperature on key enzymes involved in starch and protein formation in grains of two wheat cultivars. *J. Agro. & Crop Sci.,* **194** : 47-54.

2

RICE

ORYZA SATIVA (L.)

Family : Poaceae

Genus : Oryza

Species : Sativa

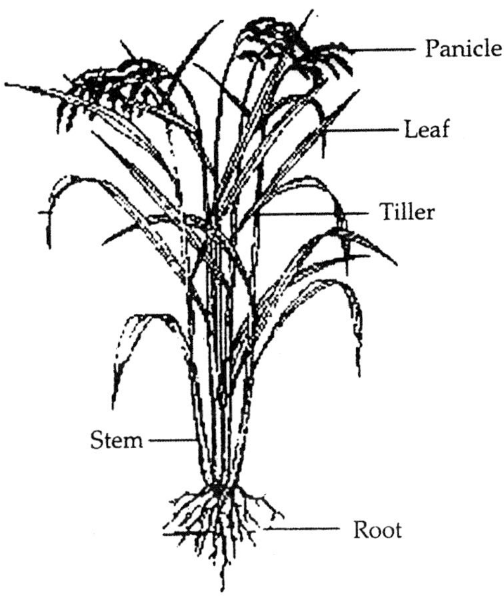

Fig. 1 : Rice plant sowing root tiller and panicle.

Rice (Fig.1) is an annual plant consisting of roots, culm and leaves and having a height of 2 to 10 feet, the lower part floating in water or prostrate. It has jointed culms or stem and the panicle is terminal and diffuse, bowing when the seed is weighty. It is probably indigenous to China and certainly to India, where the wild form grows by tanks, ditches and rivers. It was early introduced into East Africa and Syria, and later into America, where it already appears as a native plant. In Europe, rice was brought into the Mediterranean basin from Syria by the Arabs in the Middle Ages, but is now grown largely only in the plain of Lombardy, and a little in Spain. In England it has been cultivated merely as a curiosity, and may be seen in the hothouses of most botanic gardens, treated as a water plant.

India is one of the richest countries in the world with tremendous diversity in rice varieties, which depends upon the weather, soil structure, characteristics and purposes. According to Dr. Richaria, a noted scientist, 4,00,000 varieties of rice existed in India during the Vedic period. Even today 2,00,000 varieties exist in India which is indeed an exceptionally high number.

In India rice is grown in east, west, south and north region. In north region it is grown once a year during the month of June-September/October as *kharif* crop while in eastern and southern regions, it is grown, two or three times a year.

Root and tillers

Rice, at germination has seminal root. All roots are initiated at the stem nodes. Root number and weight increase with increase in tiller number attaining their maximum values at the time of heading and anthesis. Root growth is more accentuated in early stages as compared to shoot growth. The absorption of nutrients such as nitrogen, potassium, sulphur and phosphorus is usually most rapid between tillering and early panicle formation (Inadu, 1967). According to Okajima (1960), emergence of root is closely correlated with nitrogen content in stem base and active emergence of root occurs only when it is more than 1%.

Tillering begins at the four to five-leaf stage. Emergence of tillers is closely linked to that of leaves. Thus, all tillerings are synchronized with the development of leaves on the main stem (Katayama, 1951). Mother stem feeds the tillers initially as far as nutrient supply is concerned but it become autotrophic when it has three leaves and four to five roots (Ishizuka and Tanaka, 1963).

Tillering capacity is one of the most important characteristics of a variety. The initiation of tiller primordia is free from the influence of environment (Yamazaki, 1960), but their emergence and development are greatly influenced by such factors as nitrogen supply, solar radiation, and temperature. The most important of these is nitrogen content of the plant (Kumura, 1956); a concentration above 3.5% is necessary for active tillering, at 2.5% tillering stops, and below 1.5% death of tillers takes place (Ishizuka and Tanaka, 1963).

Production of tillers at different growth stage, effective tillers and non effective tillers at harvest vary significantly due to variation in variety. Singh and Jain (2000) have reported that tiller production increased sharply from active tillering to panicle initiation stage and declined gradually during flowering and remained almost constant during ripening phase in all rice cultivars. Amongst various cultivars, tall statured variety, Karnal local recorded highest number of tillers/m^3 followed by semi tall cv. IR 72, Basmati 370 and lowest was in Pusa 933, Pusa 1019 and Pusa Basmati 1.

Phosphorus level is also closely correlated with tillering, no tillering taking place when the phosphorus content of the mother stem is below 0.25% (Honya, 1961). Alam *et al.* (2009) have reported that an application of 72 kg P_2O_5/ha showed better tiller production and fertility. Plants grown without added phosphorus gave the lowest tiller production.

Climatic factors are also important. The number of emerging tillers is highest when winter temperature is 15-16°C at night and 31°C during the day, but the optimum temperature for tiller development is 31°C both day and night (Tsunoda, 1964, Matsushima *et al.*, 1964b). Tiller production is affected by photoperiod. Mishra and Khan (1973) have reported that 8 hr photoperiodic treatment for 21 day at the age of 28 days significantly reduced the production of tillers. Goldman and Mitchell (1999) have also reported that tiller production was highest in 12 hr photoperiod and the minimum in 8 hr photoperiodic treatment in a semidwarf rice culture Al-Nan-Tsao under green house condition. Thus tiller production in rice is affected by low photoperiod.

Reproductive Differentiation

Reproductive growth starts at the differentiation of young panicle and panicle formation starts when all nodes have been formed and internodes begin to elongate. Normally panicle forms 3 to 4 weeks before they are visible in the field and it takes 25 to 33 days from the internode elongation until panicle emerges from the plant.

Flowering begins at the tip of the panicle and moves progressing downward to the panicle base. Flowering which occurs from mid-morning to shortly after noon normally lasts from 6 to 10 days. Flowering however is severely affected, delayed or stopped during cloudy, rainy or cool weather.

Photosynthesis

Rice is a C_3 plant and has comparatively less photosynthetic CO_2 fixation than C_4 plants. Although efforts are being made to transfer C_4 genes of photosynthesis to rice plant, however a significant positive result is still awaited. A wide variability in photosynthetic rate in rice genotypes has been reported (Janardhan *et al.*, 1983, Basuchandhuri and Dasgupata, 1987, Sharma and Singh, 1994). A high rate of photosynthesis is associated with higher productivity, unless sink capacity is limiting (Evans, 1976). Analysis of data of correlation coefficient between photosynthetic rate, yield component and grain yield amongst rice varieties indicate that there are significant positive correlation between spikelets per m^2, grain per m^2 and the photosynthetic rate (Table 1).

Table 1 : Correlation Coefficient and Regression Equations Between Photosynthetic Rate (at flowering), Grain Yield and Yield Components

Parameters	*Correlation coefficient (r)*	*Regression equation*
Tillers per m^2	0.229^{NS}	-
Days to 50% flower	0.114^{NS}	-
Panicles per m^2	0.283^{NS}	-
Spikelets per m^2	0.652**	$Y_1 = 33.1 + 0.0434\ X$
Grains per m^2	0.186^{NS}	-
1000-grain weight (g)	-0.119^{NS}	-
Grain yield (g per m^2)	0.502**	$Y_2 = 521.3 + 0.373\ X$

NS= Not significant;

** Significant at 1%

X = Photosynthetic rate at flowering; Y_1 = Spikelets per m^2 (10^3); Y_2 = Grain per m^2 (g)

Source : Basuchaudhury and Dasgupta (1987).

The photosynthetic rate in rice vary in japonica and indica cultivars. Generally the rate of photosynthesis per unit leaf area in japonica rice varied with cultivars by around 20% from average. Ohno (1976) had reported relatively great difference in the rate of photosynthesis in indica cultivars when grown at IRRI, some of them being higher in rate of photosynthesis than the others by 30%. It was also reported that some indica rices showed higher net assimilation rates than japonica ones, and that the partial contribution of net photosynthetic rate to dry matter production was about 30%, whereas that of the leaves was about 70% of the total dry matter production (Murata, 1961). It is interesting whether or not any difference in the rate of photosynthesis of leaves exists between japonica and indica rice cultivars.

Photosynthetic rates of rice varieties of India has been examined at different stages of crop growth (Taniyama *et al.*, 1988). It has been shown that the chlorophyll content in the leaves affected significantly the rate of CO_2 uptake. Phosotynthetic activity of the leaf differed not only with the cultivar but also with its position on the culm. A close correlation was found between the chlorophyll content and the photosynthetic ability of the leaf, when pooled for all the upper three leaves of four cultivars (Vijayamashuri, HR35, Sonamahasuri and Basmati) except for the case of leaves in cv. HR59 with only a slight difference in chlorophyll content.

Fig. 2 illustrated the relationship between chlorophyll contents and the apparent photosynthesis of upper three leaves in all cultivars used. The photosynthesis clearly increased with the increase in chlorophyll content, giving a highly significant correlation coefficient. Thus, the chlorophyll content of the leaves closely related to the CO_2 uptake ability in high yielding cultivars at least, and that low rates of photosynthesis of the leaves obtained with HR59, a scented type, were mainly due to low chlorophyll content.

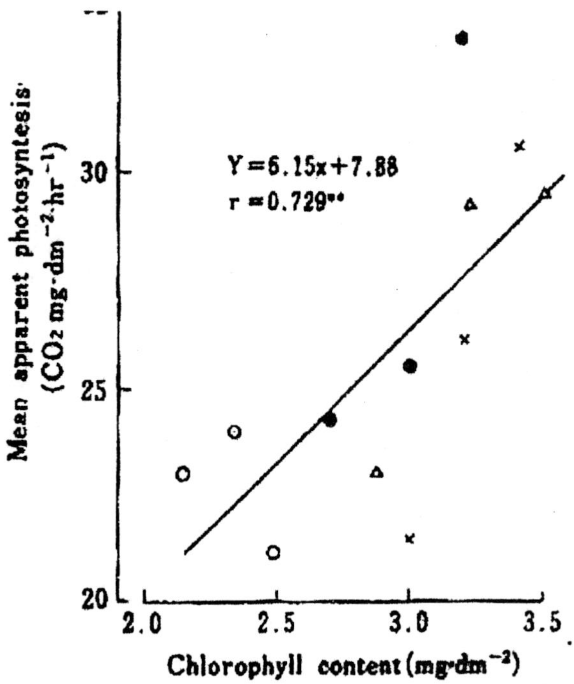

Fig. 2 : Relation between chlorophyll content and mean apparent photosynthesis Symbols Ë%,Ï%, ″ and x denote cvs. HR 59, HR 35, Vijayamahsuri and Sonamahsuri, respectively.

(*Source :* Taniyama *et al.* 1988)

The temperature and light have their effect on the photosynthetic activity in the leaf of rice cultivars (Fig. 3&4). However, there is varietal variation in response of these factors.

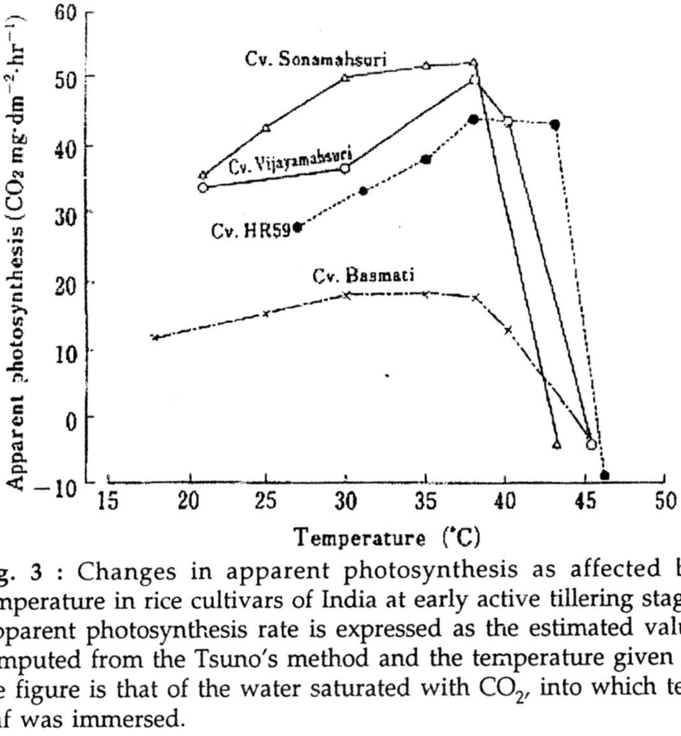

Fig. 3 : Changes in apparent photosynthesis as affected by temperature in rice cultivars of India at early active tillering stage. Apparent photosynthesis rate is expressed as the estimated value computed from the Tsuno's method and the temperature given in the figure is that of the water saturated with CO_2, into which test leaf was immersed.

(*Source* : Taniyama *et al.* 1988)

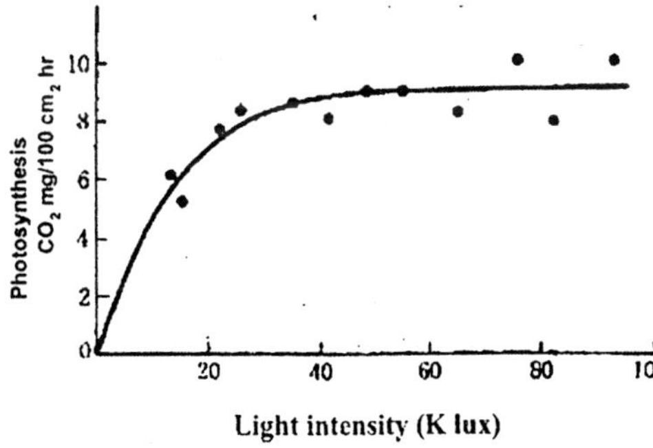

Fig. 4 : Light intensity and rate of photosynthesis per unit leaf area. (Variety: H 4) (*Source* : Akio, 1963)

In hilly regions of India, rice production is constrained by low temperature (Kumar and Mahajan, 2007). Symptom of chilling injury is visible on photosynthetic capacity along with poor germination, stunted growth, discoloration and necrosis, poor panicle exertion, and abnormal ripening etc. Photosynthesis in rice increases with increase in light intensity and reaches a saturation level (Akio, 1963).

Environmental Factors

In India, rice is grown under widely varying conditions of altitude and climate. It extends from 8 to 35°C altitude and from sea level to as high as 3000 meters. Rice crop needs a hot and humid climate. It is best suited in regions which have high humidity, prolonged sunshine and an assured supply of water. The average temperature required throughout the life period of the crop ranges from 21 to 37°C. At the time of tillering, the crop requires a higher temperature than for growth. Temperature requirement for blooming is in the range of 26.5 to 29.5°C and at the time of ripening the temperature should be between 20-25°C. Photoperiodically, rice is a short-day plant, however, there are varieties which are non-sensitive to photoperiodic conditions. Previous investigations (Sircar, 1942, 1946) on vernalisation and photoperiodism of rice have demonstrated the effectiveness of short days in inducing earliness, increased growth rate and grain yield. In the preliminary work (Sircar, 1942) the application of 8 hours photoperiods to plants at an advanced stage of the vegetative growth has resulted in a significant increase in tiller number and an earliness of ear emergence. In subsequent work (Sircar, 1944, 1946) different photoperiods have been applied to seedlings in seed beds, as the application of the method in field practices is possible only by the treatment of seedlings before transplantation. Of the seed bed treatments for varying periods, an exposure to short days of 8 and 10 hours for 6 weeks duration induces maximum earliness and increased grain yield. Tiller number during the vegetative period is maintained at a higher level than in the control.

In Indian condition, during the rice cultivation period, light intensity has significant effect on the growth and productivity of rice. During rainy season, photosynthesis is generally affected because of reduction in light intensity and the duration brought about by the excessively cloudy weather. Murty *et al.* (1975) have reported the major cause of low yield is reduced light intensity. Laxmi Praba *et al.* (2004) have suggested that the rice crop should be exposed at least 40 sunny days, particularly at heading to obtain high fertility and optimum harvest index. This is a major constraint in rice productivity in the tropics (Yoshida, 1981). The solar radiation requirement of a rice crop differs at various growth phases (Yoshida and Parao, 1976). Shading during the vegetative growth phase slightly affects rice growth and yield, while shading during the reproductive phase has a pronounced effect on sink capacity (spikelet number per panicle and per unit land area). During ripening, shading markedly reduces grain yield because of a spurt in spikelet sterility (Yoshida, 1981).

Reduction in rice yield under low radiation has been reported by Singh (2000) which was mainly due to increased spikelet sterility and decreased number of grains per panicle. Venkateshwarlu *et al.* (1977) reported that dry matter accumulation was affected under shading at all stages which was critical from primordial initiation onwards. The expression of panicle number appears to be a function of environment, particularly the light intensity within 26 to 39°C range of mean maximum temperature. Grain panicles were also reduced by shading from primordial initiation onwards. The yields were progressively reduced with low light initiation appearing in succession at different growth stages, particularly more critical during ripening phase. It was considered that low light intensity during monsoon season was an important constraint for higher productivity. There are certain varieties which show comparatively better yield than others and they appear to be shade tolerant. To assess the effect of these adverse factors on different cultivars efforts have been made to analyze the basis of such tolerance of low light. Intensive investigation carried out at the Central Rice Research Institute, Cuttack have revealed rice varieties adopted to low light such as Swarnbrabha, Pallavi, Vijaya, Vajram & Hybrid IR 54752A, etc. Under low light, chlorophyll 'b' increases with reduction in chlorophyll 'a/b' ratio. The photosynthetic rate is maintained alongwith absorption of optimum nitrogen content. These plants also have lower senescence, low respiration and high carbohydrate translocation from shoot to developing grains.

Many weedy species of *Oryza* (*O. australiensis, O. brachyantha, O. nivara, O. longistaminata, O. barthii,* and *O. rhizomatis*) show the same photosynthesis rates as cultivated rice (Yeo, Yeo and Flowers, 1994). Experiments conducted in Arkansas also showed that red rice (*O. sativa*) has the same photosynthesis rate as cultivated rice (Stiers, 2002). However, it has been observed that in growers' fields even sparsely distributed red rice plants seem to always manage to grow above the rice canopy, and short-stature red rice plants are able to grow normally in the usual dense stand of cultivated rice. However, where the situation is reversed in terms of density, cultivated rice cannot overcome the shading effect of weedy rice. Therefore, it has been theorized that weedy rice has better tolerance to shading than cultivated rice. Experiments conducted by Stiers (2002) at Stuttgart, Arkansas, in 2000 and 2001 showed that the medium-grain Bengal and long grain Wells varieties sustained 42 and 46% reduction in photosynthesis rate, respectively, two days after being subjected to 50% shade compared with only a 38% reduction for the Stuttgart SHR rice. The shade treatments were applied 10 days after 50% anthesis. The significantly lesser effect of shading on the photosynthesis rate of red rice suggests that it is more tolerant of reduced light than cultivated rice and, thus, might be able to grow well in normal rice populations even when it emerges later than the crop, or when it is set back temporarily by herbicide application. A possible mechanism of shade adaptation could be an altered chlorophyll a/b ratio in response to shade. This was observed in Stuttgart SHR rice, which showed a ratio of 4.69 chlorophyll a:b without shade and 4.19 after 2 days of exposure to 50% shade. The chlorophyll a:b ratios of rice cultivars were not altered by shading. The reduced chlorophyll a:b ratio reflects an increase in the proportion of chlorophyll b relative to chlorophyll a. An increase in chlorophyll b per unit area allows efficient harvesting of light under reduced light conditions.

The effect of low irradiation has been observed on the large subunit (LSU) of Ribulose 1,5- bisphosphate carboxylase/ oxygenase with molecular mass of 55 kDa in a cultivar IR-20 (Viji *et al.*, 1997). Analysis of chloroplast polypeptides revealed an induction of light harvesting chlorophyll protein 2 (LHCP-2) under shade. The induction was more expressed in cultivar Co43 (a shade tolerant) than cultivar IR-20 (a susceptible type)

Nitrogen Requirement

Nitrogen is one of the three major nutrients required by plant. It is highly soluble and highly mobile therefore its deficiency can be visualized at the lower stem of the plant. In flooded soil condition, the loss of nitrogen is very high. Its deficiency leads to stunted growth, yellowing of leaves and finally reduction in grain yield.

Nitrogen to rice plant is applied in the form of ammonia. Ammonia being positively charged is adsorbed on the reduced state of soil particles and not easily leached under flooded condition. If NO_3 is applied under reduced condition of soil particles, being negatively charged it may not be adsorbed on the soil particles and may get washed away along with water. Although plants do not prefer NO_3 over ammonia since ammonia can not be stored in the cell as it increases pH of the cell and may cause injurious effect, NO_3 on the other hand can be stored in plant cell with no apparent damage to the plant cell. Ammonium uptake is facilitated by a family of integral membrane protein known as ammonium transporters found in all domains of life. In rice Li-Bao-Zhen *et al.* (2009) have reported ammonium transporter (AMT) genes which play a role in controlling the activity of transporters. Rice has better system of ammonia utilization through GS-GOGAT system (Magalhaes and Huber, 1989) and thus is able to incorporate ammonical nitrate into the organic nitrogen glutamine and glutamic acid.

The requirement of nitrogen by rice plant differs at different growth and developmental stages. Quantitatively, the requirement of nitrogen for vegetative growth is far more than that required for the reproductive development. Even during the vegetative growth, the requirement of nitrogen differs at germination, seedling, tillering and panicle initiation stages. Application of nitrogen at appropriate time and amount may therefore increase grain yield substantially. Split application of nitrogenous fertilizer will be having advantage for both, to meet the nitrogen requirement at that particular stage and secondly to avoid the denitrification of applied nitrogen from leaching losses.

Water Stress

Rice is a semi-aquatic plant and commonly grown under flooded conditions. In the world, however, about half of the rice cultivation has been done with insufficient water supply to maintain flooded conditions and rice yield is reduced to some extent by drought, a period of no rainfall or no irrigation (Heuson *et al.*, 1990, Setter *et al.*, 1993). In rice the severity of drought injury and plant response to soil moisture stress differ as to the degree of soil moisture, growth stage, plant organs and genotypes, etc. (Castillo *et al.*, 1992; Ekanayake *et al.*, 1989, O'Toole and Moya, 1978). The soil moisture stress affects the rice plant directly by reducing the growth rate and tillering. At different growth stages, yield damage under water stress was more susceptible at reproductive growth phase than vegetative growth phase, especially between the meiosis and heading stages (Ekanayake *et al.*, 1990). The soil moisture stress also affects the crop indirectly but immediately, and it causes plants to decrease the rates of photosynthesis and transpiration (Fig. 5) which results in reduced water and nutrient uptake (O'Toole and Baldia, 1982) as well as assimilation (Bunce, 1982). The plant responses and adaptation mechanism to water deficit conditions or soil moisture stress in plant have been studied for a long time. Particularly, the signal interaction between roots and shoots under soil moisture stress have been partially clarified with special reference not only to the physical plant water relationship (Kramer and Boyer, 1995), but also to the synthesis and translocation of plant hormones (Wilkinson and Davies, 1997) and metabolic products in plant organs (Chen and Kao, 1993).

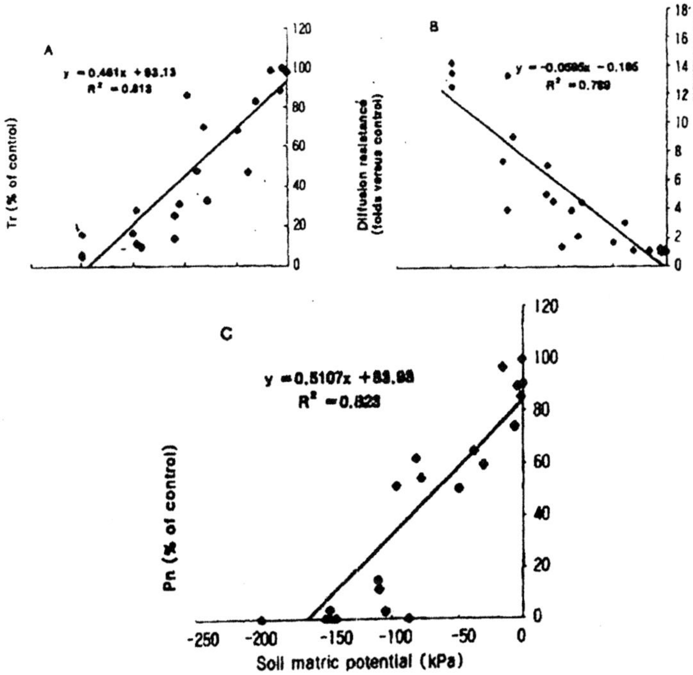

Fig. 5 : The relationships between soil matric potential and relative values of physiological parameters of the water stressed plant to the control plant; leaf transpiration rate (A), leaf diffusion resistance (B) and shoot photosynthetic rate (C). ***Source :*** Weon *et al.* (1999).

Moisture stress affects the total plant morphology and severely affects the yield. Optimum root growth is required for stabilizing shoot growth subjected to drought. A deep and thick root system has been shown to be a heritable characteristics and is essential to avoid drought. A measurable variation in root system in rice has been reported (Yoshida and Hasegawa, 1982), Kanbar *et al.* (2009) have reported that under low moisture stress, maximum root length and root number are important for improving grain yield. Thus root character may be exploited for developing a drought tolerant plant. In India not much attention has been given to root growth studies as it is a very tedious work. Leaves are the main transpiring organ in the plant which are severely affected by low moisture. Leaf rolling is the first visual symptom of drought and occurs due to inability of leaves to sustain transpiration demand of the plant (Blum, 1988). Leaf rolling reduces the leaf surface exposure to sunlight and decreases transpiration leading to closure of stomata, so that the gaseous exchange and CO_2 entry into the cells are reduced and photosynthesis is decreased. Leaf rolling is the most important criteria found useful in assessing levels of drought tolerance in large scale screening programme. (Chang *et al.*, 1974). Abd Allah (2009) has made a genetic studies on leaf rolling and some root traits under drought condition. His observation indicates that leaf rolling in susceptible rice varieties started showing rolling symptom 7 days after irrigation in the morning and evening while the resistant varieties did not roll at early morning hours but rolling starts at around 3PM as the plant transpiration demand increased after 10 days from irrigation. Plants showing leaf rolling at early stage of stress appear to have poor drought tolerance. Leaf rolling reduces photosynthetic surface and light absorption leading to reduced assimilate levels.

Several biochemical traits have been found associated with drought tolerance capacity of the rice plants. Osmoregulation and antioxidant systems have been well reported and associated with drought tolerance capacity. Many organic substances play important roles in osmoregulation including proline, glycine betaine, leaf protein, soluble sugars such as levan, trehalose, sucrose etc.

Proline accumulation has been shown to correlate with tolerance to drought and salt stress in plants. Attempts were made to compare the growth rate of transgenic rice plants in which the expression of a mothbean" 1-pyrroline-5-carboxylate synthetase (*p5cs*) cDNA was driven separately with a constitutive and a stress inducible promoter (Su and Wu, 2004). It was found that both constitutive expression and stress inducible expression of the *p5cs* cDNA in transgenic rice have led to the accumulation of *p5cs* mRNA and proline. Third generation transgenic rice seedlings showed significantly higher tolerance to stress produced by high levels of NaCl or water deficiency as judged by faster growth of shoots and roots in comparison with non transformed plants. However, stress inducible expression of the *p5cs* transgene showed significant advantages over the constitutive expression in increasing the biomass production of transgenic rice grown in soil under stress conditions (Su and Wu, 2004).

A common consequence of most abiotic stresses is an increased production of reactive oxygen species. Reactive oxygen species (ROS), viz., superoxide radical ($O_2^{\cdot -}$), hydrogen peroxide (H_2O_2) and hydroxyl radical (OH) originate from one, two or three electron transfers to dioxygen (O_2). These incompletely reduced oxygen species are toxic byproducts of various cellular oxygen consuming redox processes, such as photosynthetic or respiratory electron transport. These toxic ROS causes damage to DNA, proteins, lipids, chlorophyll and almost every other organic constituent of the living cell.

Plants protect cell and sub-cellular systems from the cytotoxic effects of these ROS with the help of antioxidant enzymes, such as superoxide dismutase, ascorbate peroxidase, glutathione reductase, mono-dehydro-ascorbate reductase, dehydro-ascorbate reductase, catalase etc. and metabolites, such as glutathione, ascorbic acid, α-tocopherol and carotenoids (Lieber *et al.*, 1986, Alocher *et al.*, 1997).

Wang *et al.* (2005) has made an attempt to develop transgenic drought tolerant rice plants with MnSOD from pea (*Pisum sativum*) under the control of an oxidative stress inducible SWPA2 promoter which was introduced into chloroplast of rice by agrobacterium mediated transformation. Functional expression of the pea MnSOD in transgenic rice plants (T_1) was revealed under drought stress induced by polyethylene glycol (PEG) 6000 (Fig.6). After PEG treatment the transgenic leaf slices showed reduced electrolyte leakage compared to wild type (WT) leaf slices, whether they were exposed to methyl viologen (MV) or not, suggesting that transgenic plants were more resistant to MVor PEG-induced oxidative stress. Transgenic plants also exhibited less injury, measured by net photosynthetic rate, when treated with PEG. It was suggested that SOD is a critical component of the ROS scavenging system in plant chloroplasts and that the expression of MnSOD can improve drought tolerance in rice.

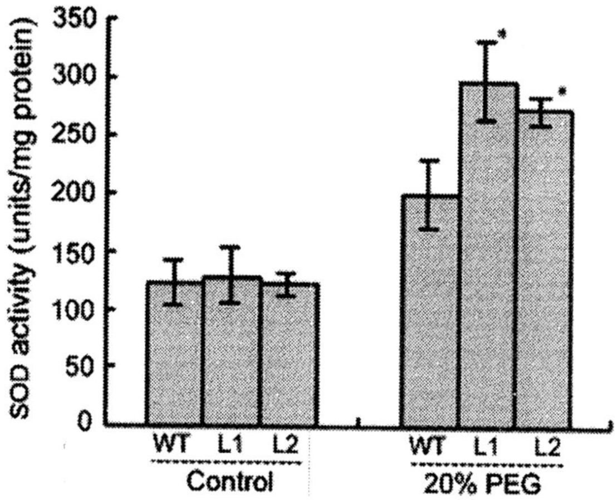

Fig. 6 : Total SOD activity (units/mg protein) in WT and T_1 transgenic plants under normal condition and after 48 h, 20% PEG treatments. L_1 & L_2 represents line 1 and 2 of T_1 trasngenic plants, respectively. Vertical bars represent standard errors (n=3 per excrement). Values significantly different from the wild type in the same group according to T-test are indicated by asterisks ($P \leq 0.05$).
Source : Wang *et al.* (2005).

Efforts have been made by other workers who have shown that H_2O_2 is involved in oxidative stress signaling leading to the induction of cytosolic ascorbate peroxidase (Morita *et al.*, 1999).

Although efforts are being made to develop the transgenic with desired gene however, a significant result on transgenic rice for drought resistance is yet to be achieved.

Flooding stress

Excess water is major constraint to productivity in most of the crop. A major constraint from excess water is an inadequately supply of oxygen to submerged tissue. Diffusion of oxygen through water is 10^4 fold slower than in normal air. In addition, excess water also leads to changes in soil which influences the phytohormone 'ethylene' and accumulation of products of anaerobic metabolisms by soil organism. When flooding results in complete submergence, availability of carbon dioxide, light and oxygen to shoot decreases.

Rice by nature is tolerant to waterlogging, flash flood and associated anoxia which makes its cultivation possible under flood prone area. Rice can germinate without oxygen, and its seedlings can grow in submerged conditions that are partially

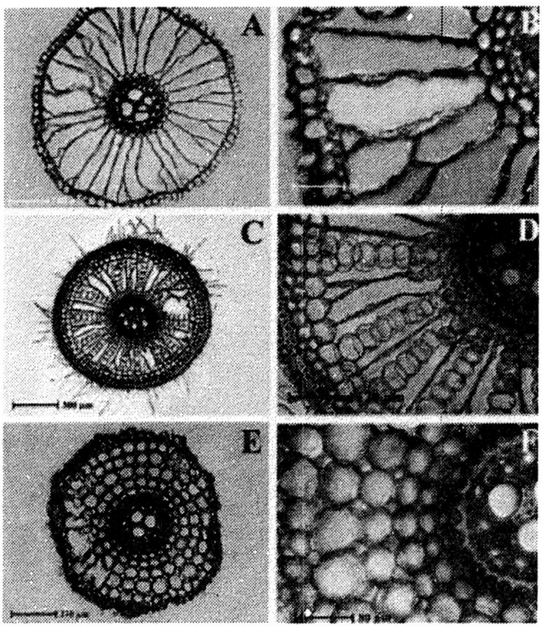

Fig. 7 : Sections of roots of rice (*Oryza sativa*) plants cultivar Zayande-Rood (A-F). Aerenchyma formation at submergence (A and B), irrigated (2 months under treatment) (C and D) and irrigated (4 months under treatment) (E and F). The cross-sections were obtained at 20 mm from the root tip.
Source : Mostajeram and Rahimi-Eichi 2008.

depleted of oxygen. The coleoptile elongates to a greater length than in normal in response to anaerobiosis, partial oxygen shortage, and enriched ethylene or carbon dioxide. These factors are inhibitory for most of the species. However, roots fail to emerge from rice seeds unless some oxygen is present, while anaerobic stimulation of coleoptile elongation is limited to only a few millimeters. Furthermore, extension of the first true leaves is dependent on oxygen, although its effect is increased by ethylene and carbon dioxide. If the shoots of small seedlings are submerged, photosynthetic fixation of carbon dioxide from the flood water is minimal and the shoot and roots depend on seed reserves for survival over at least for 3 days.

A number of morphological, physiological and biochemical characters are associated with tolerance to submergence which are influenced by genotypic, depth of submergence and its duration. Characters like coleoptile that tolerate anoxia and the ability of coleoptile to elongate vigorously when submerged provide a greater chance to escape complete inundation or minimize the duration. Other morphological characters include a root system which elongates rapidly in presence of ethylene; a highly porous morphology that facilitates internal diffusion of oxygen down the plant (Armstrong, 1971) and a hydrophobic outer surface that captures a coating of air when leaves are submerged (Fig. 7). This forms a gaseous layer that may help basipetal diffusion and mass flow of oxygen from the aerial environment while few leaves tissue remain above the water surface (Raskin and Kende, 1984).

A good amount of work has been done on biochemical basis of submergence tolerance in rice. A critical evaluation of submergence-tolerant and intolerant rice cultivars revealed that seedlings of tolerant species have 30-50% more non-structural carbohydrates (such as starch, sugar, pectins) compared to the susceptible cultivars (Setter *et al.*, 1997, Sarkar, 1998). These non-structural carbohydrates are utilized during submergence to supply the required energy for growth and maintenance metabolism (Huang *et. al.*, 2005).

Manipulation of elongation growth upon submergence have shown beneficial effect. For example, application of gibberellic acid before submergence resulted with more elongation causing reduction in plant survival whereas when GA biosynthetic inhibitor paclobutragol was used, elongation was reduced that resulted in better survival percentage

(Das *et al.,* 2005). Enhanced production of ethylene during submergence also leads to stem elongation and application of ethylene biosynthesis inhibitor 1MCP (1-methyl cyclopropene) led to the reduction in elongation and better survival (Ela *et al.,* 2003). Reducing GA/ethylene biosynthesis through biotechnological manipulation can be a useful approach to develop a submergence tolerant plant.

During anoxia, the plants produce energy through glycolysis

$$\text{Hexose} + 2\text{ADP} + 2\text{Pi} + 2\text{NAD}^+ \rightarrow 2\text{ pyruvate} + 2\text{ATP} + 2\text{NADH} + 2\text{H}^+$$

$$2\text{ pyruvate} + 2\text{NADH} + 2\text{H} \rightarrow 2\text{ Ethenol} + 2\text{NAD}^+$$

In this process the purpose of ethanol synthesis is to enable energy production to continue from glycolysis by recycling the essential cofactor NAD. The production of energy from alcoholic fermentation per mol hexose is only 2 mol of ATP, compared with 24-36 mol ATP produced in aerated tissues mainly by oxidative phosphorylation.

Pyruvate is first decarboxylated by enzyme pyruvate decarboxylase (PDC) yielding acetaldehyde which is reduced to ethanol by alcohol dehydrogenase.

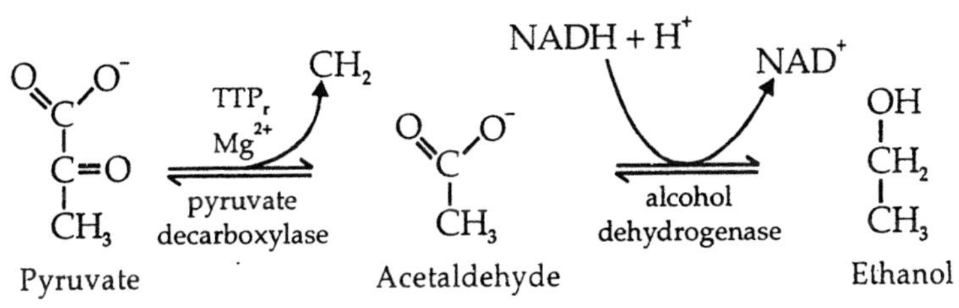

The activities of these two enzymes PDC and ADH are known to increase during submergence leading to the production of ethanol and regeneration of NAD$^+$. Ethanol concentration, however, do not reach toxic level in submergence rice seedlings (Ellis and Setter, 1999), though acetaldehyde may be toxic if its level exceeds certain limit.

Genetic manipulation of PDC and ADH levels on submergence tolerance in rice (Dennis *et al.,* 2000, Rahaman *et. al.,* 2001) have been taken up for tolerance. These studies indicated that PDC transgenic rice is much more susceptible to submergence than the wild type as the former accumulates higher level of acetaldehyde. Further, higher activity of ADH in engineered plants did not contribute to submergence tolerance, rather it made them more susceptible due to faster depletion of reserve carbohyderates. An optimum level of induction of this pathway is probably more associated with tolerance to provide minimum energy required for maintenance metabolism, but without excessive use of carbohydrates as the alcohol fermentation pathway is much less efficient than normal aerobic respiration (Sarkar *et al.,* 2006).

Low light intensity and hypoxia are the two key factors that limit the ability of rice plants to survive long periods of submergence, due to severe damage to old leaves and failure to develop new leaves. Moreover, the submerged plants are promptly exposed to an environment with high light and oxygen tension, leading to the formation of reactive oxygen species (ROS) such as superoxide anion, hydroxyl radical and hydrogen peroxide that if not removed, can cause severe damage to cellular organization leading to plant death. Cultivars that are more efficient in detoxifying the ROS upon exposure to air are more capable of retaining their chlorophyll, maintain growth of older leaves, regenerate new leaves relatively fast and hence sustain plant growth. Antioxidants and antioxidative enzymes both take care of the oxidation damage.

Natural occurring antioxidants are ascorbate, tocopherol, glutathione, carotenoides etc. Antioxidative enzymes such as superoxide desmutase (SOD, catalase, perosidase etc. are involved in deoxyfying the free radicals. Its would thus be desirable to screen germplasm for higher level of these antioxidants and antioxidative enzymes, for tolerance to water logging.

Salinity Stress

Alongwith drought, salinity is a major factor affecting agricultural productivity. Irrigation water typically contains some salt, which eventually build up, leading to salinization related losses in productivity. Attempts to generate plant varieties with improved salinity tolerance using selection base breeding strategies have proved largely unsuccessful.

Although it may be feasible to improve salinity stress tolerance using whole plant phenotypic or physiological selection strategies and pyramiding breeding schemes, such approaches, even those that employ marker-assisted selection, are slow and require massive screening efforts to identify specific quantitative trait loci. Attempts to engineer improved tolerance using single or multi-gene transfer of "candidate" genes offer far more rapid and promising improvements in stress tolerance. Such genetic engineering strategies, however, are limited by an incomplete understanding of stress tolerance mechanisms.

Osmoregulation is one of the unique characteristics for salinity stress tolerance in plants. Osmotic adjustment in plants subjected to salt stress occur by accumulation of high concentration of either inorganic or low molecular weight organic solutes such as sugars, organic acids, polyols and nitrogen containing compounds such as amino acids, amides, imino acids, actrine, proteins and quaternary ammonium compounds. Overproduction of proline under stress is reported in several varieties of rice, Pokhali (a salt tolerant variety of rice) accumulates 13 times more proline in shoot under salt stress than under normal condition. (IRRI Annual Report 1978-1979). Chutipaljit *et al.* (2009) have also reported a considerable increase in proline accumulation in rice when exposed to increasing level of salinity. Proline accumulation appears to be a variatal characteristics and has been found associated with change in activity of pyrroline-5 carboxylic acid reductase and proline dehydrogenase (Roy *et al.*, 1992). The activity of *P5C* reductase was stimulated in salt resistant varieties under stress environment. It has been therefore, proposed that proline or the activity of P–5–C reductase, a proline synthethesing enzyme, can be used as a marker for screening sensitive and tolerant varieties at early stage of plant growth. However, several reports indicate a negative correlation between proline accumulation and salt tolerance. It is also suggested that high proline may be possibly the indication of injury and may not confer osmo salt tolerance. Besides total flavonoid content also increased in salt stress tolerant variety of rice (Chutipaljit *et al.*, 2009).

Polyamine metabolism has been shown to be related to salinity tolerance in rice, however, the information is very limited. Krishnamurty and Bhagwat (1989) have studied in a number of rice varieties (salt sensitive and tolerant) the metabolism of proline with salinity tolerance. Salt sensitivity in rice was found to be associated with excess accumulation of putrescine and low levels of spermidine in the shoot system of salt sensitive cultivars. It has been suggested that one of the possible mechanism of salt resistance is highly increased level of polyamines against the low increase in diamines. Alternatively, the salt sensitivity could be due to high increase in diamines and an incapacity to maintain high level of polyamines.

A study on adoptive mechanism for cytosolic sodium homeostasis indicates that under salt stress the ability to reduce Na^+ influx into the cytosol, and subsequently increase the compartmentalization of cytosolic Na^+ in the vacuole, appeared to be the significant salt tolerance determinant in the salt tolerant rice such as pokhali and these mechanisms were either absent or less efficient in salt sensitive rice such as BRRI Dhan 29 (Kader, 2006). A lower Na^+ uptake in cv. Pokhali compared to cv. BRRI Dhan 29, depended on different types of transporters involved in these two cultivars. In cv. BRRI Dhan 29, the transporters mediating Na^+ influx were K^+ selective channels and NSCCS, whereas NSCCS were the main pathways for Na^+ uptake in cv Pokkali. Apart from lower Na^+ uptake in cv. Pokhali seem to take up Na^+ only transiently and exclude it mainly into the vacuole. Moreover to maintain a low cytosolic Na^+/K^+ ratio, Pokkali also might have increased K uptake under salt stress.

Temperature Stress

Temperature both very low or very high affects the growth and yield of rice plants. Very low temperature during germination delays rate of germination and subsequent phenology of the plant. Rice varieties can be characterized by a base temperature below which development stops (Tbase), an optimal temperature (Topt) at which development rate is fastest and above which this rate remains relatively constant and a temperature sum (Tsum) expressed in degree-days °Cd) which represents a total number of thermal units necessary for the growth and development through the physiological processes. Singh *et al.* (2010) have reported that high temperature (+2.5°C) during the vegetative and reproductive growth phases caused greater and almost equal reduction in biomass and grain yield compared with that during the ripening growth phase which showed 8% and 7% reduction in biomass and grain yield, respectively. Among yield components, the number of panicles m^{-2} and grains per panicle showed greater sensitivity to high temperature stress where 1000 grain weight was least affected by the same level of heat stress. The marked reduction in grain yield of rice under high temperature stress during vegetative growth was mainly attributed to the significant reduction in number of panicle m^{-2} followed by number of grains per panicle. The 1000 grain weight increased to a lesser extent. The reduction in grain yield by high temperature during reproductive growth phase was caused by a significant decrease in number of panicle m^{-2} and the number of grains per panicle. However, the relatively low reduction in grain yield due to heat stress during ripening was possibly caused by small reduction in 1000 grain weight and the number of grains per panicle, probably due to greater respiratory loss of assimilates in panicle.

High temperature at flowering can induce floret sterility and can limit grain yield. The main cause of floret sterility induced by high temperature at flowering is anther indehiscence (Satake and Yoshida, 1978, Matsui *et al.*, 1997a,b, 2001). High temperature at flowering inhibits swelling of the pollen grains (Matsui *et al.*, 2000b) which is the driving force behind anther dehiscence in rice (Matsui *et al.*, 1999a,b). Anthers of high temperature tolerant cultivars dehisce more easily than those of susceptible cultivars and contribute to pollination under high temperature conditions (Matsui *et al.*, 2000b, 2001). However, the factors determining the degree of anther dehiscence are unknown.

Plant ideotype concept

Donald (1968) proposed the ideotype approach to plant breeding in contrast to the empirical breeding approach of defect elimination and selection for yield per se. He defined "crop ideotype" as an idealized plant type with a specific combination of characteristics favorable for photosynthesis, growth, and grain production based on knowledge of plant and crop physiology and morphology. He argued that it would be more efficient to define a plant type that was theoretically efficient and then breed for this (Hamblin, 1993).

During the 1960s, the yield potential of the irrigated rice crop increased from 6 to 10 t/ha in the tropics. This was accomplished at IRRI, primarily by reducing plant height through the incorporation of a recessive gene, *sd1*, for short stature from Chinese variety Dee-geo-woo-gen. A good success has been achieved by scientists at International Rice Research Institute Philippines. In 1962, IRRI plant breeders made crosses that resulted in the development and release of IR8 in 1966, the first semi dwarf variety in the tropics. IR8 represented a new plant architecture-short stature, high tillering, sturdy stems, and dark green and erect leaves (Jennings 1964). The plant type represented by IR8 and succeeding varieties was extremely effective in increasing the productivity of irrigated rice lands (Khush 1995).

Plant breeders and physiologists at IRRI in the late 1980s postulated that the plant type of these varieties may limit further improvement in yield potential. They produce a large number of unproductive tillers and have excessive leaf area that may cause mutual shading and a reduction in canopy photosynthesis and sink size, especially when grown under direct-seeded conditions (Dingkuhn *et al.*, 1991). Most of these varieties have high tillering capacity and small panicles. The large number of unproductive tillers, which limit sink size and contribute to lodging susceptibility, was identified as the major constraint to yield improvement in these varieties. Furthermore, simulation models predicted that a 25% increase in yield potential was possible by modifying the following traits of the semidwarf plant type (Dingkuhn *et al.*, 1991): (1) enhanced leaf growth in combination with reduced tillering during early vegetative growth, (2) reduced leaf growth along with sustained high foliar N concentration during the reproductive stage and late vegetative growth, (3) a steeper slope of vertical N concentration gradient in the leaf canopy with more N present in the top, (4) expanded storage capacity of the stems, and (5) improved reproductive sink capacity along with an extended grain-filling period.

These factors prompted IRRI scientists to propose modifications to the high-yielding indica plant type in the late 1980s in order to increase the yield potential. The proposed new plant type (NPT) has low tillering capacity (3-4 tillers when direct seeded, 8-10 tillers when transplanted); few unproductive tillers; 200-250 grains per panicle; a plant height of 90-100 cm; thick and sturdy stems; leaves that are thick, dark green, and erect; vigorous root system; 100-130-d growth duration; and increased harvest index (Peng *et al.*, 1994, Khush, 1995). This ideotype became the "new plant type" highlighted in IRRI's strategic plan (IRRI 1989). The goal was to develop an NPT with yield potential 20-25% higher than that of existing semidwarf varieties of rice in the tropical environment during the dry season.

However, at the time, the ideotype approach to plant breeding was not considered novel. Meanwhile, IRRI has also been looking at hybrid rice breeding to increase the yield potential of rice in the tropics. Ideotype breeding and exploiting the genetic variability require a team effort-physiologists and agronomists give their inputs, while breeders do the gene juggling in the field.

Exploitation of hybrid vigour through hybrid rice breeding is another approach for increasing the yield potential of rice in the tropics. Evidence supports the merits of transferring yield-enhancing loci from exotic germplasm and wild species and efforts are under way to increase the yield potential of tropical rice. Biotechnological approaches such as the development of C_4 rice are likewise being investigated although it is still at infancy stage.

REFERENCES

Abd Allah, A.A. (2009). Genetic studies on leaf rolling and some root traits under drought conditions in rice (*Oryza sativum* L.). *African J. Biotechnology,* **8**(22): 6241-6248.

Akiko Kozaki and Takeba, G.O. (1996). Photorespiration protects C_3 plants from photo oxidation. *Nature,* **384** : 557-560.

Akio, O. (1963). Studies on the photosynthesis in rice. *Proc. Crop Society of Japan,* **33**: 69-76.

Alam, M.M., Mirza Hasanuzzamam and Nahar Kamrun (2009). Tiller dynamics of three irrigated rice varieties under varying phosphorus levels. *American-Eurasian Journal of Agronomy,* **2**(2) : 89-94.

Alocher, R.G., Donahue, J. and Cramer, L.L. (1997). Reactive oxygen species and antioxidants relationship in green cells. *Physiol. Plant,* **100** : 224-233.

Anonymous (1989). Research Highlight, Division of Plant Physiology, IARI, New Delhi Ed. Y.P. Aborl and S.N. Bhardwaj pp. 60.

Armstrong, W. (1971). Radical oxygen losses from intact rice roots as affected by distance from the apex; respiration and waterlogging. *Physiol. Plant,* **25** : 192-197.

Bao-Zhen, L.I., Merrick, Mike, Su-mei, L.I., Hong-ying, L.I., Shu-wen, Z.H.U., Wei-ming, S.H.I. and Yan-hua, S.U. (2009). Molecular basis and regulation of ammonium transporter in rice. *Rice Science,* **16**(4): 314-322.

Basuchandhari, P. and Dassgupta, D.K. (1987). Photosynthetic rate and yield relationship in some late duration high yielding rice. *Indian J. Pl. Physiol.,* **30**(4): 400-403.

.Blum, A. (1988). Plant breeding for stress environments, CRC Inc. Florida, USA.

Bunce, J.A. (1982). Effect of water stress on photosynthesis in relation to diurnal accumulation of carbohydrates in source leaves. *Can J. Bot.,* **60** : 195-200

Castillo, E.G., Buresh, R.J. and Ingram, K.T. (1992). Lowland rice yield as affected by timing of water deficit and nitrogen fertilization. *Agron. J.,* **84** : 152-159.

Chang, T.T., Loresto, G.C. and Tagum, P.O. (1974). Screening rice germplams for drought resistance, *SABRAO, J.,* **6**(1) : 9-16.

Chen, C.T. and Kao, C.H. (1993). Osmotic stress and water stress have opposite effects on pubescence and praline production in excised rice leaves. *Plant Growth Reg.,* **13** : 197-202.

Chutipaljit, S., Cha-UM, S. And Sompornpailin, K. (2009). Different accumulation of proline and flavonoids in indica rice varieties against salinity. *Pak. J. Bot.*, **41**(5) : 2497-2506.

Das, K.K., Sarkar, R.K. and Ismail, A.M. (2005). Elongation ability and non structural carbohydrate levels in relation to submergence tolerance in rice. *Plant Sci.*, **168** : 131-136.

Dennis, E, Dolferus, R; Ellis, M; Rahman, M; Wu, Y; Hoeren, F.U; Grover, A; Ismond, K.P.; Good, A and Peacock, W.J. (2000). Molecular strategies for improving waterlogging tolerance in plants. *J. Exp. Bot.*, **51** : 89-97.

Dingkuhn, M., Penning de Veies, F.W.T., De Datta, S.K., Van Laar, H.H. (1991). Concept for a new plant type for direct seeded flooded tropical rice. In Direct Seeded Flooded Rice in the Tropics. Los Banos (Philippines) International Rice Research Institute, pp. 17-38.

Donalde,C.M. (1968). The breeding of crop ideotypes. *Euphytica*, **17** : 385-403.

Ekanayake, I.J., De Datta, S.K. and Steponkus, P.L. (1989). Spikelet sterility and flowering response of rice to water stress at anthesis. *Annuals of Botany*, **63** : 257-264.

Ekanayake, I.J., Steponkus, P.L., De Datta, S.K. (1990). Sensitivity of pollination to water deficit at anthesis in upland rice. *Crop Sci.*, **30** : 310-315.

Ella, E.S., Kawano, N., Yamauchi, Y., Tanaka, K. and Ismail, A.M. (2003). Blocking ethylene perception enhances flooding tolerance in rice seedlings. *Funct. Plant Biol.*, **30** : 813-819.

Ellis, M.H. and Setter, T.L. (1999). Hypoxia induces anoxia tolerance in completely submerged rice seedlings. *J. Plant Physiol.*, **19** : 57-61.

Evans, L.T. (1975). The physiological basis of crop yield. In: Crop Physiology, Ed. L.T. Evans, Cambridge University Press, London.

Goldman, K.R. and Mitchell, C.A. (1999). Transfer from long to short photoperiod affects production efficiency of day neutral rice. *Hort. Sci.*, **34** : 822-838.

Hamblin, J. (1993). The ideotype concept: useful or outdated? In: International Crop Science I. Madison, WI (USA).

Hamdani, A.R. (1979). Low temperature problems and cold tolerance research activities for rice in India IRRI (1979). A report of rice cold tolerance workshop Los Banos Laguana Philippines PO Box 933 Manila, Philippines.

Heuson, A.D., Peakock, W.J., Evans, L.T., Arntzen, C.J. and Khush, G.S. (1990). Drought resistance in rice. *Nature*, **395** : 26-27.

Honya, K. (1961). Studies on the improvement of rice plant cultivation in volcanic ash paddy field in Tohokudistrict. *Tohoku. Agr. Expt. Sta.Bull.*, **21** : 1-143.

Huang, C., He, W., Guo, J., Chang, X., Su, P.a nd Zhang, L. (2005). Increased sensitivity to salt stress in an ascorbate deficient *Arabidopsis* mutant. *J. Exp. Bot.*, **56** : 3041-3049.

Inadu, K. (1967). Physiological characteristics of rice roots especially with the view point of plant growth stage and root age. *Bull. Nat. Inst. Agri. Sci. Japan Ser D* **16** : 19-156.

IRRI (1978). Annual report for 1977 pp. 548. Los Bans, Philippines.

IRRI, (International Rice Research Institute) (1989). IRRI. Towards 2000 and beyond. Los Banos Philippines IRRI P. 36-37.

Ishizuka, Y. and Tanaka, A. (1963). Studies on the nutrio-physiology of the rice plants. Yokendo, Tokyo, 307 pp.

Janardhan, K.V., Murty, K.S. and Rawlo, S. (1983). Path coefficient analysis and heritability of photosynthetic rate and associated leaf characteristics in rice varieties. *Ind. J. Pl. Phy.*, **26** : 168-176.

Jennings, P.R. (1964). Plant type as a rice breeding objective. *Crop Sci.*, **4** : 13-15.

Kader, Md Abdul (2006). "Salt stress in rice" doctoral diss. Deptt. of Plant Biology and Forest Genetics. *SLU Acta Universitatis Agriculturae Sceciae*, pp. 57.

Kanbar, A., Toorchi, M., and Shashidhar, H.E. (2009). Relationship between root and yield morphological characters in rainfed low land rice (*Oryza sativa* L.). *Cereal. Res. Commun.*, **37**(2) : 261-268.

Katayama, T. (1951). Studies on the tillering of rice, wheat and barley. Yokendo, Tokyo, 117pp.

Khush, G.S. (1995). Breading the yield frontier of rice. *GeoJournal*, **35** : 329-332.

Khush, G.S., Peng, S. (1996). Breading the yield frontier of rice. In: Increasing yield potential in wheat: breading the barriers. Reynolds MP, Rajaram S., McNab, A. editors, El Batan (Mexico): International Maize and Wheat Improvement Center, pp. 36-51.

Khush, G.S., Virk, P.S., Evangelista, A., Romena, B., Pamplona, A., Lopena, V., Dela, Cruz, N., Peng, S., Cruz, C.V., Cohen, M. (2001). Germplasm with high yield potential. In: 2001 annual report. Los Banos (Philippines): International Rice Research Institute, pp. 4-5.

Kramer, P.J. and Boyer, J.S. (1995). Plant water relationship, plants effect on soil moisture. In : Water Relation of Plant and Soils. *Acad. Press.*, 408-481.

Krishnamurthy, R. and Bhagwat, K.A. (1989). Polyamines as modulators of salt tolerance in rice cultivars. *Plant Physiol.*, **91** : 500-504.

Kumar Pramod and Mahajan, V. (2007). Effect of low temperature stress on photosynthesis, total soluble sugars grain filling rate and yield in rice (*Oryza sativa* L.). *Ind. J. Pl. Phy.*, **12**(3) : 253-260.

Kumura, A. (1956). Studies on the effect of internal nitrogen concentration of rice plants on the constitutional factor of yield. *Proc. Crop Sci. Soc. Japan*, **24** : 177-180.

Laxmi Praba, M., Vanangamudi and Thandapani, V. (2004). Effect of low light on yield and physiological attributes of rice. *IRRN*, **29**(2) : 71-72.

Li Bao-Zhen, Mike Merrick, Li Su-Mei, Li Hong-Ying, Zhu Shu-Wen, Shi Wei-min, Su Yan-hua (2009). Molecular basis and regulation of ammonium transporter in rice. *Rice Science*, **16**(4) : 314-322.

Lieber, D.C., King, D.S. and Reed, D.J. (1986). Antioxidant protection of phospholipids bilayers by a tocopherol control of and tocopherol status and lipid peroxidation by ascorbic acid and glutathione. *J. Biot. Chem.*, **261** : 12114-12119.

Magalhaes, J.R. and Huber, D.M. (1989). Ammonium assimilate in difficult plant species as affected by nitrogen form and pH control in solution cultivar. *Fertilizer Research*, **21**(1) : 1-6.

Matsui, T., Namuco, O.S., Ziska, L.H. and Horie, T. (1997b). Effect of high temperature and CO_2 concentration on spikelet sterility in indica rice. *Field Crop Res.*, **51** : 213-219.

Matsui, T., Omasa, K. and Horie, T. (1997a). High temperature induced spikelet sterility of japonica rice at flowering in relation to air temperature, humidity and wind velocity condition. *Japanese J. Crop Sci.*, **66** : 449-455.

Matsui, T., Omasa, K. and Horie, T. (1999a). Rapid swelling of pollen grains in response to floret opening unfolds locule in rice. *Plant Production Sci.*, **2** : 196-199.

Matsui, T., Omasa, K. and Horie, T. (1999b). Mechanism of anther dehiscence in rice (*Oryza sativa* L.). *Annals of Botany*, **84**: 501-506.

Matsui, T., Omasa, K. and Horie, T. (2000b). High temperature at flowering inhibit swelling of pollen grains, a driving forçe for the case dehiscence in rice (*Oryza sativa* L.). *Plant Production Sci.*, **3** : 430-434.

Matsui, T., Omasa, K. and Horie, T. (2001). The difference in sterility due to high temperatures during the flowering period among japonica rice varieties. *Plant Production Sci.*, **4** : 90-93.

Matsushima, S., Tanaka, T. and Hoshino, T. (1964b). Analysis of developmental factors determining yield and its application to yield prediction and culture improvement of low land rice. *Proc. Crop Sci. Soc. Japan,* **33** : 53-58.

Misra, G. and Khan, P.A. (1973). Impact of photoperiod on seasonal sowing in rice. Proceedings *Plant Sciences,* **77**(4) : 137-147.

Mostajeran,A and Rahimi-Echi,V (2008).Effect of drought stress on root anatomical characterstics of rice cultivars. Pak. J. Biol. Sci. 11 : 2173-2183

Murta, Y. (1961). Studies on the photosynthesis of rice plants and its culture significance. *Bull. Natl. Inst. Agr. Sci.,* **D9** : 1-169.

Murty, K.S., Naik, S.K. and Sahu, G. (1975). Effect of low light stress on rice crop. In: Proceedings of the Symposium on Cop Plant Response to Environmental Stresses. *Almora VPKAS,* pp. 74-84.

O'Toole, J.C. and Baloha, F.P. (1982). Water deficit and mineral uptake in rice. *Crop Sci.,* **22**: 1144-1150.

O'Toole, J.C. and Moya, T.B. (1978). Genotypic variation in maintenance of leaf water potential in rice. *Crop Sci.,* **18** : 873-874.

Ohno, Y. (1976). Variatal differences of photosynthetic efficiency and dry matter production in indica rice. *Tech. Bull. TARC,* **9** : 1-72.

Okajima, H. (1960). Studies on physiological function of the root system in the rice plant viewed from nitrogen nutrition. *Tohoku Univ. Inst. Agr. Res. Bull.,* **12** : 1-146.

Osada, Akio (1963). Studies on the photosynthesis in Indica rice. *Proceedings of the Crop Science Society of Japan,* **331** : 69-76.

Peng, S., Khush, G.S. and Cassman, K.G. (1994). Evaluation of a new plant ideotype for increased yield potential. In: Cassman K.G. Editor. Breaking the yield barrier. Proceeding of a Workshop on rice yield potential in favourable environments. Los Banos (Philippines). International Rice Research Institute, pp. 5-20.

Rahaman, M., Grover, A., Peacock, W.J., Dennis, E.S. and Ellis, M.H. (2001). Effect of manipulation of pyruvate decarboxylase and......

Raskin, I. and Kende, H. (1984). Regulation of growth in stem section of deep water rice. *Planta,* **160** : 66-72.

Roy, D., Buria, A., Basu, N. and Banerjee, S.K. (1992). Effect of NaCl salinity on metabolism of proline in salt sensitive and salt resistant cultivars of rice. *Biol. Plant,* **34** : 159-162.

Sarkar, R.K. (1998). Saccharide content and growth parameters in relation with flooding tolerance in rice. *Biol. Plant,* **40** : 597-603.

Sarkar, R.K., De, R.N., Reddy, J.N. and Ramakrishnayya, G. (1996). Studies on submergence tolerance mechanism in relation to carbohydrate, chlorophyll and specific leaf weight in rice. *J. Plant Physiol.*, **149** : 623-625.

Sarkar, R.K., Reddy, J.N., Sharma, S.G. and Ismael, A.M. (2006). Physiological basis of submergence tolerance in rice and implications for crop improvement. *Curr. Sci.*, **91**(7) : 899-906.

Satake, T. and Yoshida, S. (1978). High temperature induced sterility in indica rice at flowering. *Japanese J. Crop Sci.*, **47** : 6-10.

Setter, T.L., Ellis, M., Laurels, E.V., Ella, S., Senadhira, D., Mishra, S.B., Sarkarung, S. and Datta, S. (1997). Physiology and genetics of submergence tolerance in rice. *Ann. Bot. (Suppl. A.)*, **79** : 67-77.

Setter, T.L., Ingram, K.T. and Tuong, T.P. (1993). Environmental characterization requirements for strategic research in rice grown under adverse condition of drought, flooding or salinity. Pages 3-18. In: Rainfed Low land rice-agriculture research for high risk environments. K.T. Ingramed IRRI Los Banos Philippines.

Sharma, A.P. and Singh, S.P. (1994). Genotypic variation in photosynthesis and yield components in rice. *Ind. J. Pl. Phy.*, **37** : 188-189.

Singh, S. (2000). Growth, yield and biochemical response of rice genotypes to low light and high temperature-humidity stress. *Oryza*, **37**(1) : 35-38.

Singh, S. and Jain, M.C. (2000). Growth and yield response of traditional tall and improved semidwarf rice cultivars to moderate and high nitrogen, phosphorus and potassium levels. *Indian J. Pl. Physiol.*, **5**(1) : 38-46.

Singh, S., Aggarwal, P.K. and Yadav, R.N. (2010). Growth and yield response of rice under heat stress during vegetative, reproductive and ripening growth stages. *International Rice Res. News*, **117** : 4185.

Singh, S.D. (2005). Effect of low light stress at various growth phases on yield and yield components of two rice cultivars. *Internationl. Rice Res. Notes*, **30**(2) : 36-37.

Sircar, S.M. (1942). Studies in the physiology of rice II photosynthetic response in one variety of winter paddy. *J. Ind. Bot. Sco.*, **21** : 41-50.

Sircar, S.M. (1944). Vernalization of rice by short days. *Nature*, pp. 153-378.

Sircar, S.M. (1946). Studies in the physiology of rice III. Vernalisation by short days. *Proce. Nat. Inst. Sci. (India)*, **12** : 191-198.

Stiers, E.N. (2002). Determination of competitive characteristics in red rice (*Oryza sativa* L.). University of Arkansas, USA (M.Sc. Thesis).

Su Jin and Wu Ray (2004). Stress inducible synthesis of prolin in transgenic rice confer faster growth under stress condition than that with constitutive synthesis. *Plant Sci.,* **166**(4) : 941-943.

Taniyama, T., Subbaiah, S.V. and Rao, M.L.N. (1988). Cultivation and ecophysiology of rice plants in the tropics. III. Photosynthesis of rice cultivars of India, measured by the Tsuno simple method. *Jap. J. Crop Sci.,* **57**(1) : 184-190.

Tsunoda, S. (1964). Leaf characters and nitrogen response. In: The mineral nutrition of rice plants IRRI Johns Hopkins Press Baltimore, Maryland pp. 401-418.

Venkatishwarlu, B. Pradad, V.V.S.S. and Rao, A.V. (1977). Effect of low high intensity on different growth phases in rice (*Oryza sativa* L.). *Plant and Soil,* **47** : 37-47.

Viji, M.M., Thangaray, M. and Jayapragasam, B. (1997). Low irradiative stress tolerance in rice (*Orya sativa* L.) *Biol. Plant,* **39**(2) : 251-256.

Wang, F.Z., Wang, Q.B., Kwon, S.Y., Swak, S.S. and Su, W.A. (2005). Enhanced drought tolerance of transgenic rice plants expressing a pea manganese super oxide dismutase. *J. Pl. Physiol.,* **162**(2) : 465-472.

Weon, Young Choi, Park Hong Kyu, Kang Yong Si, Kim Su Sang and Choi Sun Young (1999). Effect of water stress on physiological traits at various growth stages of rice. *Korean J. Crop. Sci.,* **43**(3) : 282-287.

Wilkinson, S. and Davies, W.J. (1997). Xylem sap pH increase a drought signal received at the apoplastic face of the guard cell that involves the suppression of saturable abscisic acid uptake by the epidermal symplast. *Plant Physiol.,* **113** : 359-573.

Yamazaki, K. (1960). Studies on the morphogenesis of crop plants under different growing conditions. II. *Proc. Crop Sci. Soc. Japan,* **28** : 262-265.

Yeo, M.E., Yeo, A.K. and Filowers, T.J. (1994). Photosynthesis and photorespiration in the *genus oryza*. *J. Exp. Bot.,* **45**(274) : 553-560.

Yoshida, S. (1981). Fundamentals of rice crop science. Los Banos (Philippines) : International Rice Research Institute. 269p.

Yoshida, S. and Hasegawa, S. (1982). The rice root system-its development and function. Page 97-114. In: Drought resistance in crops with emphasis on rice. International Rice Research Institute, Philippines.

Yoshida, S., Parao, F.T. (1976). Climatic influence on yield and yield components of lowland rice in the tropics. In: Climate and ice. Los Banos (Philippines): International Rice Research Institute. Pp. 471-494.

3

CHICKPEA (*CICER ARIETINUM* L.)

Family	:	Fabaceae
Sub family	:	Faboideae
Genus	:	*Cicer*
Species	:	*arietinum*

Fig. 1 : Chickpea crop in field.

The chickpea is known by its different names like Bengal gram in English and Chana in Hindi. The name chickpea appears to have come from the French chickpea to Latin *cicer* (from which the Roman Cognomen (Cicero was taken). The dictionary cites 'chickpea' in the mid-18th century; the original word in English was chick, found in print in English in 1388 and taken directly from French (*Source:* USDA Nutrient database. Chickpea from Wikipedia, the free encyclopedia).

HISTORY

It is believed that species originated in the southern Caucasus and Northern Persia (Van der Maesen, 1972). Ladizinsky (1975) reported the centre of origin to be southern Turkey. Van der Maesen (1987) recognized the south east part of Turkey adjoining Syria as the possible centre of origin of chickpea based on the presence of closely related annual species *C. reticulatum*. Botanical and archaeological evidence show that chickpeas were first domesticated in Middle East and were widely cultivated in India, Mediterranean area, the Middle East and Ethiopia since antiquity. Brought to the new word, it is now important in Mexico, Argentina, Chile, Peru and the US and Australia. Wild species are most abundant in Turkey, Iran, Afghanistan and Central Asia (Duke, 1981).

BOTANY

Chickpea plants (Fig. 1) are branched, erect or spreading sometimes shrubby; 0.2 to 1 meter tall, dark green or bluish green in colors. Stem is generally grayish in appearance. The stem is branched with granular hairs on it. The main branch in gram usually produces not more than one secondary shoot, but in some types the main branches may produce numerous lateral branches.

The leaves are pinnately compound, usually with one terminal leaflet. The number as well as the size of the leaflet, however, varies in different types. There are 9-15 pairs of leaflets. The leaflets of the pinnate leaves are small, and have serrated edges. The leaves are covered with grandular hairs. The color of the leaves also varies; some being light green while other is green or dark green. Certain types possess leaflets with red margins.

Fig. 2 : Root of chickpea

Chickpea has a well developed root system upto 2 m long. The roots usually include a central strong tap root, with numerous lateral branches that spread out in all directional in the upper layer of soils but major portion exists upto 60 cm.This makes the plant to extract water from deeper layer of soil. There are numerous nodules on roots, where rhizobium bacteria are present and fix up atmospheric nitrogen (Fig. 2).

The flowers are typical papilionaceous consisting of five sepals, five petals comprised of one standard, two wings and two keels, ten stamens, nine fused to form one stamina column and one free and a carpel with the style borne laterally on the ovary (Fig. 3). They are of various colors from white to shades of pink or blue. Anthesis starts between 9 am and 10 am and may continue up to 3 pm. The flowers remain open for two days, the flowering process being over early on the second day. Self pollination is the rule, but cross pollination may occur to the extent of about 5-10% by insects. The pod is about 2 cm long and usually contains two seeds. A single plant produces about 50 to 150 pods.

The seeds are spherical in shape, wrinkled and with a pointed beak. They vary a great deal in size as well as in color.

Color of seed may vary from white, light fawn, yellowish-orange, brown, dark brownish and with a little bluish tinge. The seed coat may be smooth or puckered and wrinkled. The cotyledons are thick and yellowish in color.

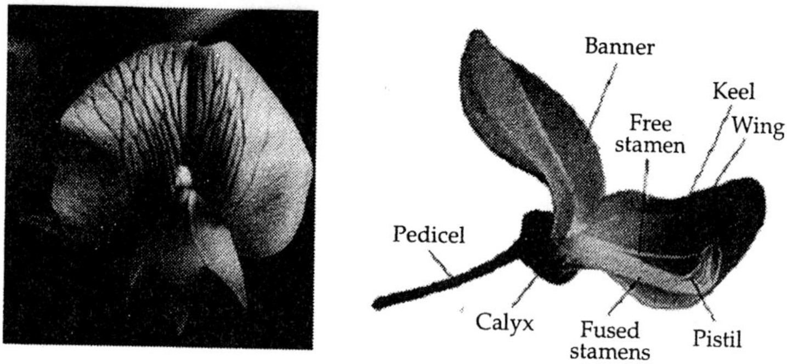

Fig. 3 : Flower and floral parts of chickpea

GENERAL CULTIVATION

Chickpea is usually grown as a rainfed cool weather crop or as a dry climate crop in semi arid region. It is valued for its nutritive seed with protein content (25 to 29%) after dehalling. Its digestibility of protein varies from 76-78% (Hulse, 1991). There are two main types of chickpea.

- *Desi,* small in size, dark with rough coat (Fig.4) cultivated in Indian subcontinent Ethiopia, Mexico and Iran.
- *Kabuli,* large in size, light colored smooth coat (Fig. 4a) cultivated in southern Europe, North Africa, Afghanistan and Chile and also in Indian subcontinent.

Fig. 4 : Desi chickpea

Fig. 4a : Kabuli chickpea

India is considered as the world leader in chickpea production followed by Pakistan and Turkey (Food and Agricultural Organization of the United National, Economic and Social Department, The Statistic Division). In India it is grown in *rabi* season (November-April). It is cultivated as a solo crop or mixed with barley, lathyrus, linseed, mustard, peas or wheat. It is also grown as a catch crop in sugarcane field and often as a second crop after rice. Chickpea yields 400-600 kg/ha but it has potential to yield 2000 kg/ha if the required inputs are provided optimally. The harvest index varies from 30 to 45% depending upon the variety and weather condition. Chickpea matures in 3-7 months depending upon the variety and weather condition.

GROWTH AND DEVELOPMENT

Germination in chickpea is fairly good under normal condition. Salinity and sodicity do affect the seed germination and growth of the seedling. Studies were made on germination of two types of seed of *kabuli* and *desi* types under the influence of salinity (2.6 and 4.9 dSm^{-1}). Working with two types of chickpea, *kabuli* and *desi*, Gholipoor *et al.* (2010) reported that under saline conditions, reduction in seedling growth, shoot water content, root and shoot K^+ concentration and increase in root and shoot Na^+ concentration were more severe in the *kabuli* type than in *desi* type cultivars. Considering path coefficients, increasing seedling K^+ concentration and uptake of water from the soil favoured salt-stressed seedling growth. Increasing K^+ content alleviated the deleterious effects of root Na^+ to a greater extent than that of shoot Na^+. On the other hand, a higher percentage of the decrease in seedling growth as the result of Na^+ was due to shoot K^+ deficiency than to root K^+ deficiency.

Low temperature delays the seed germination, however hydropriming for 12 hours or osmopriming (PGE-0.5 MPa) for 24 hours improves germination of chickpea under cold soil condition (Elkoca *et. al.*, 2007).

Chickpea is a C_3 plants and the rate of dry matter accumulation is slow for a long period. Beech and Leach (1989) reported growth rate in chickpea from 6.2 to 9.6 g m^{-2} d^{-1} in two subsequent cropping seasons. They further recorded that for about 100 days after sowing, the dry matter production was linearly related to GDD (growing degree days). Mean leaf area index (LAI) reached a peak of 2.62 at 133 days after sowing. The growth in leaf area which determines the proportion of radiation intercepted, has been reported to be linearly related to the mean daily temperature that the crop was exposed to during vegetative growth. The correlation between growth and accumulated GDD appeared to be better than that between phenology and GDD (Beech and Leach, 1989).

The duration of growth period is drastically affected by moisture availability. Sharma *et al.* (2007), while studying with six chickpea genotypes under soil moisture stress condition have reported that increased soil moisture stress reduced the maturity duration by 15 and 19 days under mild and severe stress, respectively over irrigated control. The shorter maturity duration resulted in reduction of reproduction phase. The plant water status got reduced significantly at full bloom stage (110 days) under moisture stress condition. Plant height, number of effective pods per plant, 100 seed weight, seed yield biological yield and harvest index all got significantly reduced over control. The longer maturity duration has significant and positive association with seed yield (r=0.93). In similar type of work, Singh *et al.* (2008) have reported that in chickpea, leaf area index had significant and positive correlation with total dry matter, total chlorophyll content, seed yield and harvest index (Table 1). Growing degree days (GDD) at flowering stage showed a significant positive correlation with primary branches, days to 50% flowering, RGR, NAR, RWC, seed yield, biological yield and significant negative correlation with membrane injury index. Further, membrane injury index had significant negative correlation with days to 50% flowering, GDD at flowering NAR, RWR, RWC, seed yield and biological yield.

In India due to previous crop such as rice, planting of chickpea is usually delayed in north western part of the country. Under these situation, the crop is sown in December. Thus the late sown crop faces, low temperature during germination and

Table 1 : Correlation Coefficient Among Different Traits in Chickpea

Characters	*1*	*2*	*3*	*4*	*5*	*6*	*7*	*8*	*9*	*10*	*11*	*12*	*13*
Primary branches	1.000	-0.192	0.785**	0.760**	-0.201	0.617**	0.642**	0.126	-0.440	0.562*	0.659**	0.733**	-0.488*
Leaf area index (%)		1.000	-0.343	-0.392	0.867**	-0.038	-0.065	0.763**	-0.365	-0.209	0.627**	-0.008	0.854**
50% flowering (days)			1.000	0.990*	-0.249	0.834**	0.829**	0.047	-0.553*	0.823**	0.116	0.680**	-0.638**
GDD at flowering				1.000	-0.318	0.783**	-0.764**	-0.058	-0.531*	0.822**	0.032	0.620**	-0.677**
Total dry wt. (g/m^2)					1.000	0.097	0.146	0.823**	-0.292	-0.238	0.699**	0.175	0.736**
RGR ($g\ g^{-1}\ day^{-1}$)						1.000	0.975**	0.368	-0.687**	0.827**	0.462*	0.730**	-0.348
NAR ($mg\ dm^{-2}\ day^{-1}$)							1.000	0.411	-0.599**	0.730**	0.486*	0.776**	-0.348
Total chl. content ($mg\ g^{-1}$ fw.)								1.000	-0.361	0.261	0.830**	0.452*	0.557*
Membrane injury index (%)									1.000	-0.745**	-0.454*	-0.454*	-0.034
RWC (%)										1.000	0.502*	0.487*	-0.468*
Seed yield (g/plant)											1.000	0.693**	0.518*
Biological yield (g/plant)												1.000	-0.251
Harvest index (%)													1.000

*significant at 5% level, **significant at 1% level;

Source : Singh *et al.* (2008)

initial growth stages and higher temperature during post flowering phase advances leaving less time for biomass accumulation. A critical analysis of the physiological and morphological characters of six genotypes Pusa-256, Pusa-372, BG-72, DG-36, DG-46 and DG-51 sown at different sowing dates has revealed that there is significant difference in leaf area index (LAI) and crop growth rates (Table 2 and 3).

Table 2 : Leaf Area Index (LAI) of Chickpea Genotypes of Different Sowing Dates (Mean of 6 Varieties)

Date of sowing	*Days After Sowing*			
	30	*60*	*90*	*105*
Nov. 15	0.147(0.004*)	0.396(0.012*)	1.231(0.051*)	1.506(0.047*)
Nov. 30	0.089(0.004*)	0.358(0.011*)	1.424(0.049*)	1.558(0.043*)
Dec. 15	0.063(0.007*)	0.310(0.019*)	1.342(0.085*)	0.964(0.074*)

* CD at 5% in parenthesis (Source: Singh *et al.*, 2005)

Table 3 : Crop Growth Rate (g m^{-2} d^{-1}) of Chickpea Genotypes of Different Sowing Dates (Mean of 6 Varieties)

Date of sowing	*Days After Sowing*		
	30-60	*60-90*	*90-105*
Nov., 15	1.56(0.041*)	2.39(0.181*)	5.34(0.211*)
Nov., 30	1.11(0.043*)	4.68(0.190*)	4.93(0.176*)
Dec., 15	0.67(0.075*)	4.85(0.329*)	3.75(0.306*)

*CD at 5% in parenthesis (***Source*** : Singh *et al.*, 2005)

Crop development depends upon the temperature requirement. It is tough to predict plant growth based on the calendar because temperature can vary greatly from year to year. Instead growing degree days which are based on actual temperatures are a simple and accurate way to predict when a crop stage will occur. Heat units expressed in growing degree days (GDD) are frequently used to describe the timing of biological processes. In an experiment with six genotypes of chickpea including some advance breeding lines such as DG 36, DG 46 and DG 51, the accumulated heat units or GDD taken to achieve different phenolocial stages i.e. 50% flowering, 50% podding and maturity are shown in Fig 5 (Singh *et al.*, 2005). In general, the growing degree days GDD requirement for 50% flowering, 50% pod formation and maturity were higher in the first planting followed by second and third planting. Further, the advance breeding lines required less GDD compared to released varieties and were associated with faster growth and early attainment of phenophases.

Low temperature at flowering also is a major constraints to improved yield of chickpea in some parts of the Indian subcontinent and also in Australia. Lower temperature less than day temperature (< 20°C) and night temperature <10°C affects both the development and function of reproductive structure in chickpea flowers (Nayyar *et al.*, 2005). These temperature leads to floral abortion, poor seed filling and hence reduced yield. Cold induced floral abortion has been attributed to improved pollen tube growth in the style and consequent failure of fertilization (Clarke and Siddique, 2004). A tolerance to cold stress has been related with high ascorbic acid and proline content in chickpea (Kumar *et al.*, 2010)

Beech and Leach (1989) reported that biomass accumulation in chickpea is linearly related to intercepted photosynthetically active radiation (PAR). The efficiency of conversion of intercepted PAR to biomass is similar to those of sorghum and sunflower (Leech *et al.*, 1986).

Chickpea has an indeterminate growth habit, meaning flowering and pod filling will continue simultaneously or alternatively as along as temperature and moisture permits growth to occur. About 85% of chickpea is rainfed where moisture stress plays a vital role, more so when cultivars for stress tolerance have not been identified. The present day elite cultivars have failed to make an impact due to indeterminate growth habit which often results in excessive vegetative growth at the expense of reproductive growth. Thus under both the situations, rainfed as well as irrigated, there is no assurance of yield. The detailed studies have revealed that majority of the present day cultivars need around 15% moisture in the root zone (comprising 50 cm depth) during the growth period so as to produce optimum stabilized yields. The moisture levels below or above this regime drastically affect plant growth, dry matter production and yield (Anonymous, 1989).

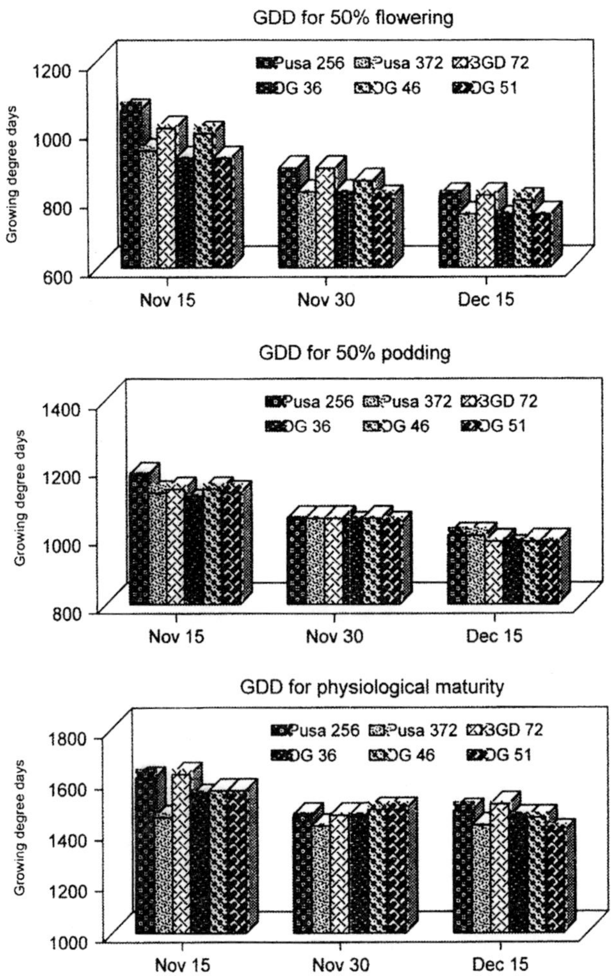

Fig. 5 : Growing degree days for different phenophases of chickpea genotypes as influenced by date of planting.

Source : Singh *et al.,* 2005.

Physiological studies (RGR), LAI, Biological yield (DM) and partitioning (HI) have provided sufficient clues which will enable the farmer to obtain better yields from the present genotypes and consequently would offset the vagaries of nature. For the rainfed/unirrigated conditions the depth of sowing is important. Compared to shallow sown crop under the same situation, deeper sowing results in saving of one irrigation. The major handicap with irrigation has been excessive vegetative growth which is caused by excessive inputs, especially moisture. This situation is very often encountered when rain follows the irrigation or irrigation is given when moisture is already optimum.

The excessive vegetative growth occurs as a result of prolonged vegetative phase or reversion of reproductive phase into vegetative in which a potential pod bearing node produces a branch. Thus, control of vegetative growth will be very useful way, till cultivars with determinate growth habit are developed.

A logical way to control vegetative growth will be to curtail the growth period through delay in watering. The detailed studies revealed that though late sowing improved HI to some extent, the dry matter production is drastically affected causing equivalent yield losses. However, it was found that if dry matter is kept constant, superior and stabilized yield could be obtained. This was achieved by increased population per unit area by narrowing the row to row distance (Baldeo, 1988).

Harvesting time is also very important for good yield. Sait Adak *et al.* (2007) reported that yield and mineral nutritional value of chickpea were significantly affected by harvest timing. When compared to the seed yield at optimal harvest time, seed yield was 18% and 9% lower in the early harvest and 27% and 31% in the late harvest in Site 1 and Site 2, respectively. Late harvest of chickpea crops resulted in significant pod dropping and shattering. Generally, protein, phosphorus (P), calcium (Ca), magnesium (Mg), copper (Cu), zinc (Zn), and manganese (Mn) concentrations of the seeds in optimal harvest were found to be greater than in early and late harvested plants. Harvest timing also results in significant variations in straw mineral nutrient concentrations of the plants. As the results of this study, it was concluded that the harvest timing is critical for yield losses and mineral nutritional value of chickpea seeds.

SOURCE/SINK

Chickpea is an indeterminant plant and flowering and pod development proceed acropetally. The excessive vegetative growth occurs as a result of prolonged vegetative phase or reversion of reproductive phase into vegetative in which a potential pod bearing node produces a branch. Sheldrek and Saxena (1979) reported that plants grown under normal field condition, at successively apical nodes of the branches, there was a decline in pod number per node, weight per pod, seed number per pod and or weight per seed. The percentage of nitrogen in the seed was same in earlier and later formed pods at Hyderabad, at Hissar the later formed seeds contained a higher percentage. Earlier and later formed flowers contained similar number of ovules. The decline in seed number per weight per seed in the later formed pods of 28 out of 29 cultivars studied, indicated that pod filling was limited by the supply of assimilates or other nutrients. However, in one of exceptivelly small seeded cultivar, there was no decline in the number of or weight of seeds in later formed pods indicating that yield was limited by 'sink size' (Sheldrake and Saxena, 1979).

Source sink relationship in chickpea was studied by removal of leaves and flowers. Pandey (1984) reported that the removal of leaves subtending flowers and pods during the reproductive phase resulted in a 40-60% reduction in seed yield compared with control plants, while removal of leaves below the flowering nodes decreased the seed yield by 25%. Complete defoliation resulted in a seed yield reduction close to 90%. The removal of leaves to reduce leaf area by 25-100% at different periods of crop growth resulted in reductions in seed yield, compete leaf removal at 100 days after planting (during early pod filling) caused the maximum reduction in seed yield but the degree of reduction was not proportional to the area of leaf removed. The number of pods per plant was the yield component most sensitive to defoliation, followed by the number of seeds per pod and the weight per seed.

The removal of 25% of the flowers throughout the reproductive phase or of all flowers and young pods upto 2 weeks after anthesis did not adversely affect seed yield. Yield compensation occurred as a result of increased pod set, number of seeds per pod and seed size. An increased intensity (50-75%) of removal of flowers and pods or removal of all flowers extending beyond 3 weeks after anthesis reduced both seed yield and total shoot dry weight.

Flower and fruit drop is one of the serious problem of chickpea, resulting in its poor reproductive efficiency viz. the number of flowers formed/retained and the per cent fruits developed with normal seeds (Van Schaik and Probst, 1958) and consequently poor yield. Ethrel, ethylene producing commercial preparation has been used to improve the yield by promoting pod growth (Abbas, 1991) and decreasing the flower and pod shedding and thereby reducing abscission and improving better pod set.

PHOTOSYNTHESIS

Chickpea belongs to C_3 group of photosynthesis. Rate of photosynthesis in leaves starts increasing from initial vegetative phase to pod formation and subsequently declined from pod filling to maturity (Singh *et al.*, 1987).

Leaf photosynthetic rate in chickpea cultivars vary between 15 and 40 mg CO_2 dm^{-2} h^{-1}. This rate compares favourable with photosynthetic rate in other C_3 crops such as wheat, barley etc. Cultivars of diverse geographic origin exhibit large difference in photosynthetic efficiency ranging between 200 and 400 µg CO_2 cm^{-1} h^{-1} (Van der Maesen 1972). Studying with 13 chickpea genotypes Mythili and Nair (1996) have reported a variability in photosynthetic activity and advocated photosynthetic rate as best selection criteria for higher productivity.

CO_2 fixation by leaves is not uniform throughout the life period. The rate decreased from flowering to pod formation. Both pods and stems contribute to total photosynthesis in chickpea (Pandey *et. al.*, 1976, 1978). Pod wall remain green for a considerable period and because they are photosynthetically active, contribute to grain yield. Differences in pod photosynthesis also exist amongst cultivars (Santakumari and Sinha, 1972). Dhawan and Singh (1983) have observed that pod wall contained chlorophyll and RuBP carboxylase activity and is capable of fixing CO_2. Although the rate of photosynthesis and RuBP case activity was higher in leaves, on fresh weight basis, the pod wall was more efficient on chlorophyll basis indicating that pod wall photosynthesis could be a source of carbohydrate for seed filling and translocation of assimilate to the nodules.

The predominant role of the pod wall in refixation of carbon is supported by measurements of fixation of isotopically labeled CO_2, which show that more than 80% of CO_2 is fixed by this tissue when provided to the pod interior. Chlorophyll concentration are of the same order for embryo, seed coat, and pod wall tissues in younger pods on both on area and a fresh weight basis, but decline differentially with development from 12-30 days after podding. Imaging of chlorophyll distribution in the pod wall suggests that less than 15% of chloroplasts are located in the inner layer of cells thought to refix CO_2 in legumes; this would be sufficient to refix less than 40% of respired CO_2. It was, therefore, concluded that while all tissues of the pod are capable of refixing respiratory carbon, the entire pod wall is responsible for the majority of this process, rather than a specialized layer of cells on the inner epidermis (Furbank *et al.*, 2004). This internal recycling of CO_2 inside the pod may assist in maintaining seed filling in water stressed chickpea (Ma *et al.*, 2000) and also under other abiotic stress environment.

Seed growth in chickpea is strongly associated with the activity of the sugar enzyme, sucrose synthase in the cotyledons with final seed weight at maturity being closely associated ($R^2 = 0.70$) with the maximum sucrose synthase activity in both irrigated and water limited conditions (Kumar and Turner, 2009). The strong association between sucrose synthase during rapid seed filling process and final dry weight accumulation suggests that sink strength is an important determinant of seed size in chickpea. Water availability during seed filling affects the sink strength as determined by sucrose synthase. The higher sucrose synthase activity in cotyledons of chickpea is considered an important trait in the breeding programme for improved seed size in chickpea irrespective of the growing environment (Kumar and Turner, 2009).

In chickpea, a high dark respiration in every part of the plant during all phases of growth has been reported by Singh (1988). Much less photosynthate was used for dry matter production than used in respiration. On an average only 17, 33 and 21% of the total photosynthate was utilized for incorporation into dry matter during vegetative, flowering and early fruiting and seed filling stages respectively. For nodule only 5-7% of photosynthate was allocated. Singh (1988) has further emphasized that for maximizing the efficiency of photosynthetic utilization for seed dry matter and protein, there is need to (a) lower the respiratory losses and delayed leaf abscission, (b) efficient mobilization of carbon and nitrogen from stem and leaves and (c) high photosynthetic and nitrogen fixing rate during the seed filling phase.

ABIOTIC STRESS

Normally, legumes are cultivated as a rainfed crop. Rabi legumes grow in the conserved soil moisture and by the time flowering and fruiting starts, the water is almost exhausted. Seed filling thus faces water stress which ultimately results in poor yield. If irrigation is provided to chickpea crop during vegetative and pod filling stages, the yield is expected to improve. However, the observation is that if irrigation level is high, there is an excessive vegetative growth resulting in poor fruit setting due to poor harvest index. Nodulation and nitrogen fixation are very sensitive to low water potential (Khanna Chopra *et al.*, 1984).

Under favorable conditions, the symbiotic N_2 fixation can produce as high as 176 kg N ha^{-1} (Beck, 1992), and provide upto 85% of its N requirements (Rennie and Dubetz, 1986). Inoculation with an effective strain of *Rhizobium* is an economical way of enhancing seed yield and quality in chickpea (El Hadi and Elsheikh, 1999). However, rhizosphere colonization and nodule formation can be reduced substantially when soil moisture is low (Paul, 1998; Gan *et al.*, 2005). Likewise, the proportion of N derived from symbiotic N_2 fixation is also decreased by water stress (Marcellos *et al.*, 1998; Kurdali *et al.*, 2002). Under drier conditions, chickpea can derive only a small proportion of crop N requirements from symbiotic N_2 fixation (Marcellos *et al.*, 1998). This is because soil moisture plays a critical role in both nodule formation and N_2 fixation. Low soil moisture during the early stages of the plant growth decreases nodule formation (Gan *et al.*, 2005) and low moisture during late vegetative to early flowering period decreases efficiency of N_2 fixation (Beck *et al.*, 1991). An early onset of terminal drought can have a severe impact on the seed yield of chickpea relying on symbiotic N_2 fixation (Wery *et al.*, 1988). In case when the N_2 fixation activity is limited by terminal drought stress, chickpea plant may receive benefits from the use of low rates of fertilizer N when some energy may be derived from the combination of N source (Munns, 1977). Studies by Gan *et al.* (2008) have shown that moisture stress had a negative effect on nodulation but this negative effect was reduced significantly when plants were supplied with low doses of fertilizer N. They further suggested that lower accumulation of amides in nodules in the presence of N resulted in less N to be transported out of the nodules. Biomass partitioning data revealed that the use of fertilizer N reduced the amount of biomass invested in nodules, but it increased biomass investment in roots. Stronger root systems allowed plants to allocate more water for the transport of fixed N to aboveground organs. In practice, application of low doses of fertilizer N increased pod fertility and seed yield of chickpea, particularly when the inoculated chickpea was under moisture stress during reproductive development. Although nodule numbers and their functionality were negatively affected by the application of inorganic N, greater seed yield with low doses of fertilizer N was achieved in chickpea.

Several approaches are being made to find the basis of stress tolerance in chickpea and to develop some selection criteria for breeding for drought resistance. For drought resistance, three mechanisms has been suggested (Turner *et al.* 2001).

A. Earliness

Most important mechanism to escape the onset of drought under terminal condition is the earliness. Short duration varieties of chickpea maturing before onset of drought have proved to be successful (Kumar *et al.*, 1996). The optimum crop duration for maximum yield depends upon water availability.

B. Drought Postponement or Avoidance

This can be achieved by maintaining water uptake and reducing water loss. In order to have this, large root system for greater extraction of water and small leaf area for reducing transpiration loss are useful.

C. Drought Tolerance

Drought tolerance means the ability of plant to tolerate the water deficit and plant can metabolise at low water potential.

There are many ways that plant can maintain leaf water potential such as by closing of the stomata by action of abscisic acid. Abscisic acid is endogenously synthesized as a result of water stress. In many plant species, ABA transgenic for stress tolerance have also been developed however their performance in field condition is yet to be ascertained.

OSMOREGULATION

Osmoregulation with the accumulation of certain compatible solutes maintain the osmotic adjustment and maintains growth of the plant. Osmotic adjustment (OA) has been suggested as an important trait in postponing dehydration in water limited environments as it maintains cell turgor and physiological processes as water deficit develops (Turner and Jones, 1980; Morgan, 1984). Variation on OA among chickpea cultivars in response to developing water deficits has been observed in several studies (Singh *et al.*, 1990; Morgan *et al.*, 1991; Lecoeur *et al.*, 1992; Leport *et al.*, 1999; Moinuddin and Khannu-Chopra, 2004). However, the relationship between OA and yield of chickpea under water limited environments is inconsistent. Morgan *et al.* (1991) indicated that the degree of

OA observed under controlled conditions was positively correlated with the yield of the cultivar under rainfed conditions in the field. Moinuddin and Khanna-Chopra (2004) found that seed yield of chickpea was correlated with the degree of OA when grown under a line-source irrigation system in the field. However, Leport *et al.* (1999), who observed a range of OA in chickpea from 0 to 1.3 MPa, did not observe any effect on yield, and Singh *et al.* (1990) observed that OA did not always result in a yield increase, particularly in cultivars that had the greatest degree of OA and partitioned a large fraction of assimilates to the root. Indeed, Serraj and Sinclair (2002) concluded that no yield advantage from OA had been demonstrated in any crop except at very low and uneconomic yields.

Turner *et al.* (2006) has analyzed the association between the degree of OA and grain yield of advanced breeding lines (ABLS) of chickpea developed from crosses between cultivars with high and low OA. The association was assessed under rainfed conditions in the field in Australia and India in F_8 progeny which were assumed to be stable for the OA trait. Their studies have confirmed that OA varies in chickpea and it has also demonstrated that the phenotypic expression of OA is not stable, but appears to vary from year to year in each genotype, suggesting that it may be the passive consequence of differences in rates of accumulation and utilization of solutes rather than an active accumulation of solutes in response to a water deficit. The study has also demonstrated that where there are differences in OA, this has no effect on the yield under terminal drought when grown on current rainfall in Mediterranean type climates and when grown on stored soil moisture in subtropical climates in India. Thus, OA cannot be considered a selectable drought resistance trait in chickpea breeding programmes.

Membrane Stability Index

It is associated with the leakage of solutes from the cell. Drought causes membrane injury and the extent of damage depends upon the severity of the stress and variety.Singh *et al* have shown a negative correlation between membrane injury index and seed yield. A selection criteria based on the membrane injury index has been suggested in many crop plants. Deshmukh and Kushwaha, (2002) have categorized the susceptible and tolerant type genotypes of chickpea growing under saline environment (Fig. 6).

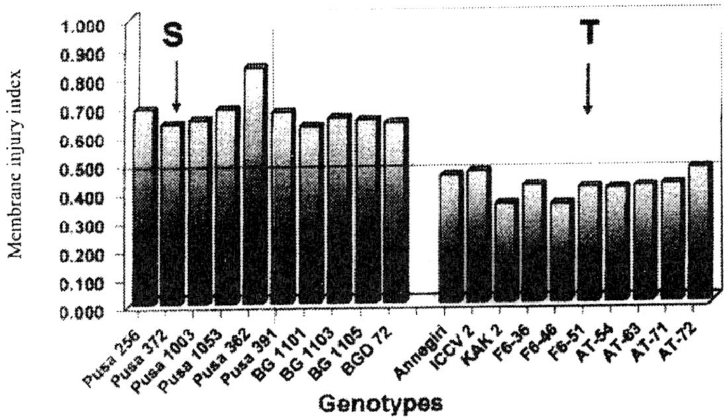

Fig. 6 : Membrane injury index at 90 DAS of different genotypes of chickpea under salinity condition at Karnal. S = Susceptible, T = Tolerant.

Source : Deshmukh and Kushwaha (2002).

Parameshwarappa and Salimath (2008) have also reported association of drought susceptibility index with membrane injury index in chickpea.

ANTIOXIDATIVE ENZYMES

It is now well established that biotic and abiotic stresses induce oxidative stress in plants. The oxidative stress caused by environmental stresses can be taken care by antioxidative enzymes such as superoxide dismutase, catalase, peroxidase, glutathione reductase etc. The ability of plant to control or reduce the oxidant levels is highly related to stress tolerance. Any stress or significant change in the environment is associated with up or down regulation of hundreds of genes and some proteins important to oxidative metabolism. In chickpea, approximately 205 protein spots were reported which are differentially regulated under dehydration. Mass spectrometry analysis allowed identification of 147 differentially expressed protein, presumably involved in a variety of functions. The dehydration responsive nuclear proteome of chickpea revealed a coordinated response which involves both the regulatory as well as functional proteins (Pandey *et al.,* 2008). Enhanced superoxide dismutase, catalase, peroxidase activities have been reported under high boron level in two chickpea cultivars (Murat *et al.,* 2009).

TEMPERATURE

In chickpea, a greater part of reproductive phase is exposed to high temperature affecting seed growth and finally yield. (Terminal flowers abort to the extent of 40-50%). Higher temperature leads to the formation of reactive oxygen species causing damage to the cell membranes. This effect is counter-acted by reactive oxygen scavengers such as antioxidative enzymes (Sairam and Saxena, 2000). In a study by Arun Kumar *et al.* (2008) in two chickpea cultivas exposed to high temperature of 35°C, a higher activity of superoxidase enzyme was reported (Table 4).

Table 4 : Effect of High Temperature Stress (35°C for 4 h at 95 DAS) on Superoxide Dismutase Activity (Units min^{-1} g^{-1} fw) in Chickpea Genotypes)

Genotypes	*Control*	*Treatment*	*Mean %*
Pusa 1103	12.58	26.81	19.69
Pusa 261	10.69	19.79	15.24
Mean	11.63	23.30	
CD at 5%			
Genotypes (G)	0.86		
Treatment (T)	0.86		
G × T	1.20		

Source : Arunkumar *et al.* (2008).

Further analysis of native PAGE revealed three SOD isoforms with form of there active bands; SOD I, SOD II and SOD III in both the genotypes. Using KCN and H_2O_2 as selective inhibitors, these isoforms were identified as Mn SOD, Cu/Zn

SOD I and Cu/Zn-SOD II. Fe-SOD is however found to be lacking in both genotypes in both control and high temperature treatments (Table 7&8).

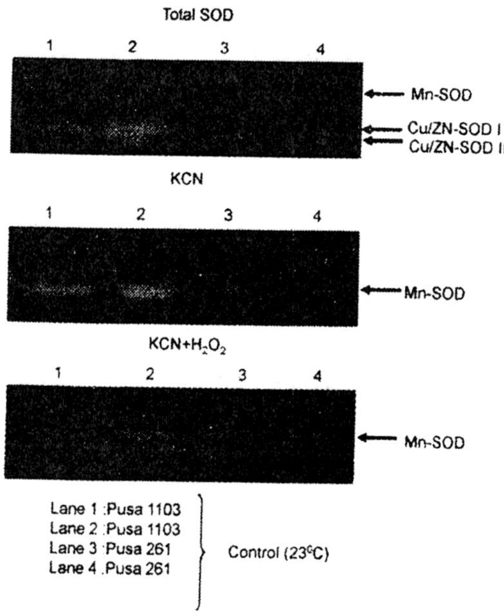

Fig. 7 : Effect of temperature (23°C, 95 days after sowing) on superoxide dismutase (SOD) isozyme patterns in chickpea genotypes. Lane 1 = Pusa 1103, Lane 2 = Pusa 1103, Lane 3 = Pusa 261, Lan 4 = Pusa 261

Chickpea is sensitive to low temperature. There is little understanding of the basic physiological and biochemical changes that occur in the plants. Temperature has a major influence on the timing of the reproductive phase (Roberts *et al.*, 1985) and secondly chilling temperature below 15°C has a deleterious effect on yield as a result of sensitivity during flowering (Savithri *et al.*, 1980; Siddique and Sedgely, 1986; Srinivasan *et al.*, 1999, Clarke and Siddique, 2004).

There has been little success world wide in breeding and selection of chickpea for tolerance to chilling (Crojer *et al.*, 2003). Screening for tolerance in field is laborious and slow process with interaction of many other environmental factors. This along with low genetic variation means that breeding chilling tolerant varieties by traditional hybridization and field screening method appears to be unsatisfactory. Clark *et al.* (2004) have suggested that pollen selection under chilling stress at hybridization as part of the breeding process could be the possible solution. Chilling temperature reduces the pollen viability, rate of pollen germination and rate of pollen tube growth both *in vitro* and *in situ* (Clarke and Siddique, 2004). It is beleived that pollen selection has a good potential for chickpea improvement since a significant variability between genotypes in response of chickpea pollen to temperature (Clarke *et. al.*, 2004).

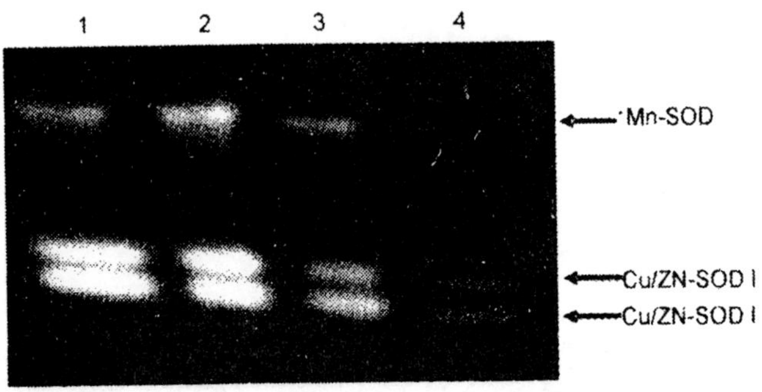

Fig. 8 : Effect of temperature (23°C for 4 h and 35°C for 6 h, 95 DAS) on superoxide dismutase (SOD) isozyme patterns in chickpea genotypes. Treatment 1 = (35°C for 4 h), Lane 1 = Pusa 1103, Lane 2 = Pusa 261, Treatment 2 (35°C for 6 h), Lane 3 = Pusa 1103, Lane 4 = Pusa 261.

Source : Arunkumar *et al.* (2008).

Besides, a thorough and a systematic studies are required to find the nature of pollen injury and the biochemical basis of tolerance such as role of antioxidative enzymes low molecular weight proteins, organic molecules including plant hormones etc. provide a stable growth of proteins at lower temperature. A combination of all these traits may provide a cold tolerant chickpea plant.

Bakht *et al.* (2006) has reported that in chickpea exposure to low non-freezing temperature and application of abscisic acid (ABA) increased the level of fatty acid desaturation in the plasma membrane (measured as the double bond index (DBI). Unsaturated lipids in the plasma membrane may prevent cell lyses at low temperature. However, they have envisaged that physical properties of the plasma membrane other than fluidity are involved in frost acclimation in chickpea.

SALINITY

Salinity is one of the important stress that significantly affects the growth and yield by inhibiting several metabolic processes. Besides, nodules which are important source of nitrogen fixation, are severely damaged. Kukreja *et al.,* (2010) have reported a progressive increase in nodular injury and reduction in RWC and osmotic potential (Ès) with increasing salinity levels in chickpea. Under the circumstances, both salinity and water deficiency cause damage to nodules. It has been further reported that lipid peroxidation goes along with ethylene formation and the decrease in the functional efficiency of chickpea nodules is due to overproduction of C_2H_4, increase in H_2O_2 content, Na^+/K^+ ratio and chlorine content along with decrease in water status and ascorbic acid content. Although antioxidant defense system was activated by increasing the activity of antioxidative enzymes, they could not overcome the accumulation of H_2O_2 in nodules.

A significant change in proline and MDA (malondialdehyde) content in leaf of chickpea was also reported by Eyidogan, and Oz (2007). In general a dose dependent decrease was observed in H_2O_2 content and photosynthesis under salt stress. It was further reported that there was a significant enhancement in activities of Cu SOD and Mn SOD along with ascorbate peroxidase and glutathione reductase enzyme activities.

Several approaches are being made to over express genes for synthesis of high level of osmoprotectants to develop transgenics in chickpea, considering Proline, (mutant gamma-pyrrolidine-5-carboxylate synthase), Mannitol (mannitol-1-phosphate dehydrogenase) and glycinebetaine (choline oxidase). Over expression of antioxidative enzymes have also been taken by many workers. However, the results are yet to come with good amount of tolerance under field condition.

PLANT TYPE CONCEPT

Seed yields of chickpea are low compared to many other crops. One of the reasons for not growing chickpea in high input conditions is that the available varieties do not respond favourably to high fertility and irrigation.. As chickpea has indeterminate growth habit, excess water promotes vegetative growth, which acts as competitive sink for developing pods and seeds (Khanna-Chopra and Sinha, 1990). Although one or two irrigations considerably enhance the yield, particularly in moisture stress conditions, a decline in yield occurs at high irrigation levels due to excessive vegetative growth of the plants and poor pod set (Sinha *et. al.*, 1985).

The concept of specific plant types, as in cereals and field peas, has received little attention in chickpeas. Headley and Ambrose (1981) suggested that increased biological yield of chickpea at higher densities tends to be compensated by reduced harvest index (HI) resulting in little change in grain yield. In high density stands of freely branching plants, branch number did not decline, but later branches competed poorly with earlier branches for resources needed to form reproductive sites and seed-forming structures. The result was that harvest index of later branches, and hence their yield, and crop yield were reduced. Siddique *et al.* (1984) proposed that seed yield in chickpea is controlled by the harvest index of branches and that seed yields might be raised by restricting branches in high density stands to no more than two, each with a high harvest index.

After testing the concept, Siddique *et al.* (1984) supported that reduced branching in high density standards should be considered. The current plant type in chickpea is unsuitable for producing high yield. Restructuring of plant type, which was a key to enhancing productivity of wheat and rice, is needed for bringing a breakthrough in chickpea productivity as well. Attempts have been made to define ideotype of chickpea for different growing conditions. Jain (1975) suggested that major gains in chickpea yield could be achieved by developing a plant type with high harvest index, response to increased plant population per unit area, and early maturity. He further suggested that improvement in harvest index would likely to be associated with determinate and compact growth habit. Bahl and Jain (1977) included erect growth habit, many primary and secondary branches and few tertiary and later order branches in chickpea ideotype. According to them this plant type would intercept more sunlight for photosynthesis and permit large population per unit area. Sedgley *et al.* (1990) also emphasized that an ideotype for high input environments should have erect growth habit and limited branching.

The wealth of genes present with wild *Cicer species* await exploitation in the breeding programme. Singh (1990) has suggested that new plant types for different situation are requited to be tailored. In this context, increased biomass, input responsive cultivar and types for high plant density situation need to be looked into. Bahl (1980) made an effort to breed for erectness with tall growth by hybridization between conventional spreading and tall types but with no success in improvement in harvest index. For improved physiological effect, Pudir and Reddy (1998) has emphases the need for smaller leaves by having fewer leaflets. Dahiya and Lather (1990) and other workers have given emphasis on short internodes and compact growth in the ideotype. Such plant type is expected to give high harvest index, accomodate more plants per unit area and resist excessive vegetative growth under high input condition. A plant type mutant with thick stem, erect, compact, dwarf and bold seeded has been identified from gram strain E100Y (M). An attempt has been made to study the inheritance and its utilization for developing good plant type in chickpea (Sandhu *et al.*, 1990). The mutant E100 YM appears to be the only source of short internode and compact growth habit. The brachytic mutant JGM-1 (Gaur *et. al.*, 2008) which was induced mutant through treatment of seed with ethyl methane sulphate (EMS) may provide alternate source for the desired traits. However, a lot of work has to be done in this direction for improving productivity in chickpea under a particular environmental conditions.

REFERENCES

Abbas, S. (1991). Biosynthesis pathways as control points in ethylene regulated flower and fruit drop and seed absorption in chickpea. Proc. Symp. Grain Legume Feb. 9-11, organized by Indian Society for Genetic & Plant Breeding, IARI, New Delhi.

Anonymous (1989). Research high lights (1997-1988). Division of Plant Physiology, Indian Agricultural Research Institute, New Delhi. p. 59.

Arun Kumar, R., Deshmukh, P.S., Singh, T.P., Kushwaha, S.R. and Bhagat, K. (2008). Superoxide dismutase in chickpea genotypes under high temperature. *Indian J. Pl. Physiol.*, **13**(1) : 88-90.

Bahl, P.N. (1980). Kabuli-desi introgression and genesis of new plant type in chickpea. In: Proceeding of the International Workshop on Chickpea Improvement, ICRISAT, Patancheru, AP, India, pp. 75-80.

Bahl, P.N. and Jain, H.K. (1977). Association among agronomic characters and plant ideotype in chickpea (*Cicer arietinum* L.). Z. *Pflanzenzuechtung.* **79** : 154-159.

Bakht, J., Bano, A. and Dominy, P. (2006). The role of abscisic acid and low temperature in chickpea (*Cicier arietinum* L.) cold tolerance II. Effect on plasma membrane structure and function. *J. Exp. Bot.*, **57**(14): 3707-3715.

Baldeo, B. (1988). Research highlights (1977-1988). Plant Physiology, IARI, New Delhi. pp. 27.

Beck D.P. (1992). Yield and nitrogen fixation of chickpea cultivars in response to inoculation with selected rhizobial strains. *Agron. J.*, **84** : 510-516.

Beck, D.P., Wery, J., Saxena, M.C. and Ayadi, A. (1991). Dinitrogen fixation and nitrogen balance in cool-season food legumes. *Agron. J.*, **83** : 334-341.

Beech, D.F. and Leach, G.J. (1989). Effect of plant density and row spacing on the yield of chickpea (cv. Tyson) grown on the Darling Downs, south-eastern Queensland. *Aust. J. Exp. Agric.*, **29** : 241-246.

Clarke, H., Siddique, K.H.M. (2004). Response of chickpea genotypes to low temperature stress during reproductive development. *Field Crops Res.* (in press).

Clarke, H.J. and Siddique, K.H.M. (2009). Response of chickpea genotypes to low temperature stress during reproductive development. *Field Crop Res.*, **90**(2-3) : 323-334.

Clarke, H.J., Khan, T.N. and Siddique, K.H.M. (2004). Pollen selection for chilling tolerance at hybridization leads to improved chickpea cultivars. *Euphytica,* **139** : 65-74.

Crojer, J.S., Clarke, H.J., Siddique, K.H.M. and Khan, T.N. (2003). Low temperature stress. Implication for chickpea (*Cicer aeritinum* L.) improvement. *Crit. Rev. Plant Sci.,* **22** : 185-219.

Dahiya, B.S. and Lather, V.S. (1990). A break through in chickpea yield paper 1. *International Chickpea News.* **22** : 6-8.

Deshmukh,P.S and Kushwaha S.R (2002).Variability in membrane injury index in chickpea genotypes.Indian J.Pl. Physiol 7(3) 285-287.

Dhawan, R. and Singh, R. (1983). Relative photosynthetic rates of leaves and pods of chickpea (*Cicer aeritinum* L.) cultivars differing in seed weight. *Indian J. Pl. Physiol.,* **26** : 276-284.

Duke, J.A. (1981). Handbook of Legumes of World. Economic Importance Plenum Press, New York, pp. 52-57.

El Hadi, E.A. and Elsheikh, E.A.E. (1999). Effect of *Rhizobium* inoculation and nitrogen fertilization on yield and protein content of six chickpea (*Cicer arietinum* L.) cultivars in marginal soils under irrigation. *Nutr. Cycl. Agroecosyst.* **54** : 57-63.

Elkoca, E., Haliloglu, K., Esitken, A. and Ercisli, S. (2007). Hydro and osmopruning improve chickpea germination. *Acta Agriculturae Scandinavica- Plant Soil Sciences,* **57**(3) : 193-200.

Eyidogan, F. and Oz, M.T. (2007). Effect of salinity on antioxidant responses of chickpea seedlings. *Acta physiol. Plant,* **29** : L 485-493.

Furbank, R.T., White, R., Palta, J.A. and Turner, N.C. (2004). Internal recycling of respiratory CO_2 in pods of chickpea (*Cicere arietinum* L.): the role of pod wall, seed coat, and embryo. *J. Exp. Bot.,* **55**(403): 1687-1696.

Gan, Y.T., Jayakumar, P., Symons, S. and McDonald, C.L. (2008). Synergic effect of N and moisture on biochemical property of nodules and seed yield in chickpea. *Australian J. Crop Sci.,* **1**(1) : 11-22.

Gan, Y.T., Selles, F., Hansen, K.G., Zentner, R.P., McKonkey, B.G. and McDonald, C.L. (2005). Effect of formulation and placement of *Mesorhizobium* inoculants for chickpea in the semiarid Canadian Prairies. *Can. J. Plant Sci.,* **85** : 555-560.

Gaur, P.M., Gour, V.K. and Sirinivasan, S. (2008). An induced brachytic mutant of chickpea and its possible use in ideotype breeding. *Euphytica,* **159** : 35-41.

Gholipoor, M., Golezani, K.G., Khooie, F.R. and Moghaddam, M. (2010). Effect of salinity on initial seedling growth of chickpea (*Cicere arietinum* L.). *Acta Agronomica Hungarica,* **48**(4) : 337-343.

Headley, C.L. and Ambrose, M.J. (1981). Designing leafless plants for improving yield of the dried pea crop. *Adv. Agron.*, **34** : 225-227.

Hulse, J.H. (1991). Nature, composition and utilization of grain legume pp. 11-27. In use of tropical legumes : Proceedings of a consultants meeting 27-30 March, 1989. ICRISAT Centre, ICRIGHT Patancheru, AP. 502324, India.

Jain, H.K. (1975). Breeding for yield and other attributes in grain legume. *Indian J. Genet.*, **35** : 169-187.

Khanna-Chopra, R. and Sinha, S.K. (1990). Wheat limits the yield of pulses. Plant Processes or plant type. In: Sinha, S.K., Sane, P.V., Bhargava, S.C. and Agrawal, P.K. (eds.). Proceedings International Congress in Plant Physiology, New Delhi. Vol. 1. pp. 268-278.

Khanna-Chopra, R., Koundal, R.K. and Sinha, S.K. (1984). A simple technique is studying water deficit effect on nitrogen fixation in nodules without influencing the whole plant. *Plant Physiol.*, **76** : 254-256.

Kukreja, S, Nandwal, A.S., Kumar, N., Singh, S., Sharma, S.K., Devi, S. and Kumar, A. (2010). Ethylene evolution and modification of antioxidant defense mechanism as indices of salinity stress tolerance in *Cicer arietinum* L. nodules. *Indian J. Plant Physiol.*, **15**(3) : 203-212.

Kumar, A. and Turner, N.C. (2009). Growth and sucrose synthase activity of developing chickpea (*Cicer aeritinum* L.) seeds under field condition. *Aust. J. Crop. Sci.*, **3**(1) : 20-27.

Kumar, J., Sethi, S.C., Johansen, C., Kelly, T.G., Rahman, M.M. and Van-Rheenerr, H.A. (1996). Potential of short duration chickpea varieties. *J. Dryland Agric. Res. Dev.*, **11** : 28-32.

Kumar, S., Malik, J., Thakur, P., Kaistha, S., Sharma, K.D., Upadhyay, H.D., Berger, J.D. and Nayyar, H. (2010). Growth and metabolic responses of contrasting chickpea (*Cicer arietinum* L.) genotypes to chilling at reproduction phase. *Acta Physiologiae Plantarum*, **10** : 1007.

Kurdali, F., Al-Ain, F. and Al-Shamma, M. (2002). Nodulation, dry matter production, and N_2 fixation by fababean and chickpea as affected by soil moisture and potassium fertilizer. *J. Plant Nutr.*, **25** : 355-368.

Ladizinsky, G. (1975). A new cicer from Turkey. *Notes from Royal Botanic Garden of Edinburgh*, **34** : 201-202.

Leach, G.J., Rees, M.C. and Charles, D.A. (1986). Relation between summer crops and ground cover legumes in a subtropical environment II. Effect of sorghum and sunflower on the growth of *Vigna trilobata* and *Medicago scutellata*. *Field Crop Res.*, **15**(1) : 39-55.

Lecoeur, J., Wery, J. and Turc, O. (1992). Osmotic adjustment as a mechanism of dehydration postponement in chickpea (*Cicer arietinum* L.) leaves. *Plant and Soil*, **144** : 177-189.

Leport, L., Turner, N.C., French, R.J., Barr, M.D., Duda, R., Davies, S.L., Tennant, D. and Siddique, K.H.M. (1999). Physiological responses of chickpea genotypes to terminal drought in a Mediterranean-type environment. *European J. Agron.*, **11** : 279-291.

Ma, Q., Behboudian, M.H., Turner, N.C. and Palta, J.A. (2000). Gas exchange by pods and subtending leaves and internal recycling of CO_2 by pods of chickpea (*Cicere arietinum* L.) subjected to water deficit. *J. Exp. Bot.*, **52**(354) : 123-137.

Marcellos, H., Fenton, W.L. and Herridge, D.F. (1998). Chickpea in wheat-based cropping systems of northern New South Wales I. N_2 fixation and influence of soil nitrate and water. *Aust. J. Agric. Res.*, **49** : 391-400.

Moinuddin and Khanna-Chopra, R. (2004). Osmotic adjustment in chickpea in relation to seed yield and yield parameters. *Crop Sci.*, **44** : 449-455.

Morgan, J.M. (1984). Osmoregulation and water stress in higher plants. *Annual Review of Plant Physiology*, **35** : 299-319.

Morgan, J.M., Rodriguez-Maribona, B. and Knights, E.J. (1991). Adaptation to water deficits in chickpea breeding lines by osmoregulation: relationship to grain yield in the field. *Field Crops Res.*, **27** : 61-70.

Munns, D.N. (1977). Mineral nutrition and the legume symbiosis. Pages 353-391. **In** : Hardy, R.W.F. and Gibson, A.H. (Eds.). A treatise on dinitrogen fixation. Sect. IV. Agronomy and Ecology. John Wiley and Sons, NY.

Murat, A., Hediye, S.A., Ismail, T., Suleyman, T. and Filiz, O. (2009). Effect of boron toxicity on root antioxidant systems of two chickpea (*Ciciere arietinum* L.) cultivars. *Plant and Soil*, **314** : 99-108.

Mythili, J.B. and Nair, T.V.R. (1996). Relationship between photosynthetic carbon exchange rate, specific leaf mass and other leaf characteristics in chickpea genotypes. *Aust. J. Pl. Physiol.*, **23** : 617-622.

Nayyar, H., Bains, T. and Kumar, S. (2005). Low temperature induced floral abortion in chickpea relationship to abscisic acid and cryprotectant in reproductive organs. *Environmental and Experimental Botany*, **53** : 39-47.

Pandey, A., Chakraborty, S., Datta, A. and Chakraborty, N. (2008). Proteomics approach to identify dehydration responsive nuclear protein from chickpea (*Cicer aeritinum* L.). *Mol. Cell Proteomics*, **7** : 88-107.

Pandey, R.K. (1984). Influence source and sink removal on seed yield of chickpea (*Cicere arietinum* L.). *Field Crop Res.*, **8** : 159-168.

Pandey, R.K.; Saxena, M.C.; Singh, V.B. and Singh, K.B. (1978). Assimilation and utilization of 14C assimilates n chickpea. J. Nuclear Agri. Biol. 7(2) 56-60.

Pandey, R.K. Saxena, M.C.; Kulabarme, M.H. Singh, V.B. and Prasad VVS (1976). Genotypic variation in photosynthetic rate and respiratory losses in some grain legumes.Plant Bioch.J. 3 : 72-80.

Parameshwarappa, S.G. and Salimath, P.M. (2008). Field screening of chickpea genotypes for drought resistance. *Karnataka J. Agric. Sci.*, **21**(1) : 113-114.

Paul, S. (1998). Symbiosis between mungbean and rhizobia: Influence of water stress. *Indian J. Plant Physiol.*, **3** : 11-16.

Pundir, R.P.S. and Reddy, G.V. (1998). Two new traits-open flower and small leaf in chickpea (*Cicer arietnium* L.). *Euphytica*, **102** : 357-361.

Rennie, R.J. and Dubetz, S. (1986). Nitrogen-15-determined nitrogen fixation in field-grown chickpea, lentil, fababean, and field pea. *Agron. J.*, **78** : 654-660.

Roberts, E.H., Hadley, P. and Summerfield, R.J. (1985). Effects of temperature and photoperiod on flowering in chickpea. *Ann. Bot.*, **55** : 881-892.

Sairam, R.K. and Saxena, D.C. (2000). Oxidative stress and antioxidants in wheat genotypes. Possible mechanism of water stress tolerance. *J. Agron. Crop Sci.*, **184**(1) : 55-61.

Sait Adak, M., Kayan, N., Gunes, A., Inal, A., Alpaslan, M., Cicek, N. and Guzelordu, T. (2007). Effect of harvest timing on yield and mineral nutritional value of *kabuli* type chickpea. *J. Pl. Nutrition*, **30**(9) : 1397-1407.

Sandhu, J.S., Brar, H.S. and Verma, M.M. (1990). Inheritance of a plant type mutant and its utilization in chickpea. *Euphytica*, **48** : 111-112.

Santakumari, P. and Sinha, S.K. (1972). Variation in chlorophyll and photosynthetic rates of cultivars of Bengal gram (*Cicer arietinum* L.). *Indian J. Exp. Biol.*, **13**(2) : 208-209.

Savithri, K.S., Ganapathy, P.S. and Sinha, S.K. (1980). Sensitivity to low tolerance in pollen germination and fruit set in *Cicer arietinum* L. *J. Exp. Bot.*, **31** : 475-481

Sedgley, R.H., Siddique, K.H.M. and Walton, G.H. (1990). Chickpea ideotypes for Mediterranean environments. In: van Rheenen, H., Saxena, M.C., Walby, B.J., Hall, S.D. (eds.), Chickpea in the nineties: proceedings for the second international workshop on chickpea improvement. ICRI-SAT, Patancheru, AP, India, pp. 87-91.

Serraj, R. and Sinclair, T.R. (2002). Osmolyte accumulation: can it really help increase crop yield under drought conditions? *Plant, Cell and Environ.*, **25** : 333-341.

Sharma, K.D., Pannu, R.K., Tyafi, P.K., Chaudhary, B.D. and Singh, D.P. (2007). Water use efficiency and yield of chickpea genotypes as influenced by soil moisture availability. *Indian J. Pl. Physiol.*, **12**(2) : 168-172.

Sheldrake, A.R. and Saxena, N.P. (1978). Comparison of earlier and late formed pods of chickpea (*Cicer aeritinum* L.). *Ann. Bot.*, **43** : 467-473.

Sheldrake, A.R. and Saxena, N.P. (1979). Growth and development of chickpea under progressive moisture stress. In Stress Physiology in Crop Plant. Eds. Mussell, H. and Staples, R.C. pp. 466-483, John Wiley and Sons, Chickpea, UK.

Siddieque, K.H.M., Sedglay, R.H. and Harshall, C. (1984). Effect of plant density on growth and harvest index in chickpea. *Field Crop Abst.*, **9** : 193-203.

Siddique, K.H.M. and Sedgley, R.H. (1985). Evaluating a reduced branching ideotype of chickpea for low rainfall Mediterranean environments. *International Chickpea Newsletter*, **13** : 14-15.

Siddique, K.H.M. and Sedgley, R.H. (1986). Chickpea (*Cicer aeritinum* L.), a potential grain legume for South-Western Australia: Seasonal growth and yield. *Aust. J. Agric. Res.*, **37** : 245-261.

Singh, D.P., Chaudhary, B.D., Singh, P., Sharma, H.C. and Karwasra, S.P.S. (1990). Drought tolerance in oilseed *Brassica* and chickpea . Hisar, India: Haryana Agricultural University, pp. 1-60.

Singh, D.P., Singh, P., Sharma, H.C. and Turner, N.C. (1987). Influence of water deficit on the water relations, canopy gas exchange and yield of chickpea (*Cicer aeritinum* L.). *Field Crop Res.*, **16** : 231-241.

Singh, K.B. (1990). Prospect of developing new genetic material and breeding methodology for chickpea improvement. *Options Mediterra-necnnes-Serie Seminaires*, N°9, 43-50.

Singh, R. (1988). Assimilation and partitioning of carbon and nitrogen during growth and development of grain legumes. Proceedings of the International Congress of Plant Physiology, New Delhi. *Soci. Pl. Phy. & Bioch.*, **1** : 261-267.

Singh, T.P., Deshmukh, P.S. and Kumar, P. (2008). Relationship between physiological traits in chickpea (*Cicere arietinum* L.). under rainfed condition. *Indian J. Pl. Physiol.*, **13**(4) : 411-413.

Singh, T.P., Deshmukh, P.S., Srivastava, G.C., Kushwaha, S.R. and Mishra, S.K. (2005). Growth rate of chickpea (*Cicer aeritinum* L. genotypes under different planting dates. *Indian J. Pl. Physiol.*, **10**(3): 254-259.

Sinha, S.K., Aggarwal, P.K. and Khanna-Chopra, R. (1985). Phenological and physiological irrigation in India. *Adv. Irrigation*, **3** : 130-212.

Srinivasan, A., Saxena, N.P. and Johansen, C. (1999). Cold tolerance during early reproductive growth of chickpea (*Cicer aeritinum* L.). Genetic variation in gamete development and function. *Field Crops Res.*, **60** : 209-222.

Turner, N.C. and Jones, M.M. (1980). Turgor maintenance by osmotic adjustment. **In** : Turner, N.C. and Kramer, P.J. (Eds.). Adaptation of Plants to Water and High Temperature Stress. New York: Wiley, pp. 87-103.

Turner, N.C., Abbo, S., Berger, J.D., Chaturvedi, S.K., French, R.J., Ludwig, C., Mannur, D.M., Singh, S.J. and Yadava, H.S. (2006). Osmotic adjustment in chickpea (*Cicer arietinum* L.) results in no yield benefit under terminal drought. *J. Exp. Bot.*, **6** : 1-8.

Turner, N.C., Wright, G.C. and Siddique, K.H.M. (2001). Adaptation of grain legumes (pulses) to water limited environments. *Advances in Agron.*, **71** : 193-231.

Van der Maesen, L.J.G. (1972). *Cicer* L. A monograph of the genus with special reference to the chickpea (*Cicer aeritinum* L.). Its ecology and cultivation. Commun. Agric. University, Wageningen Dordrecht, The Netherlands.

Van der Maesen, L.J.G. (1987). Cicer L. Origin, history and taxonomy of chickpea. Pp. 11-34. In: M.C. Saxena and K.B. Singh (ed.). The chickpea CAB International Cambrian News Ltd. Aberystwyth, UK.

Van Schaik, P.H. and Probst, A.H. (1958). Effect of some environmental factors on flowers production and reproductive efficiency. *Agron. J.*, pp. 192-197.

Wery, J., Deschamps, M. and Leger-Cresson, N. (1988). Influence of some agro-climatic factors and agronomic practices on nitrogen nutrition of chickpea. Pages 287-301 **In** : Beck, D. and Materon, L.A. (Eds.) Nitrogen fixation by legumes in Mediterranean agriculture ICARDA-Martinus Nijhoff.

4

SUNFLOWER

Class	:	Asterids
Order	:	Asterales
Family	:	Asteraceae/Compositeae
Sub Family	:	Asteroideae
Genus	:	*Helianthus*
Species	:	*annuus*

Sunflower belongs to family compositeae, genus *Helianthus* and species *annuus*. The latin name for sunflower is *Helianthus* which derives from the name Helios who was the Greek god of the sun. The sunflix-anthos means flower. The sunflower's method of tracking the sun is paralleled in Greek mythology. Helios was drowned by the Titans but he then rose and became the sun (Houret, 2002).

GENERAL INTRODUCTION

Sunflower (Fig. 1) is native of Americas. It probably originated in Mexico as far back as 8000 years ago. In America, it was grown by Red Indians for food. In the middle of 16th century it was introduced to Europe. Sunflowers were primarily used as a decorative plant until the mid of 1900 when Russians began to cultivate it as an oil crop. At present it is grown in many countries including Canada, Hungry, Yugoslavia, Bulgaria, Romania, Chilli. In India it was introduced as an oil crop in late sixtees with the introduction of five varieties namely EC 68413, EC 68414, EC 68415, EC 69874 and Sunrise selection. It is now grown in Madhya Pradesh, Maharashtra, Gujarat, Rajasthan, Mysore, Tamil Nadu during *kharif* (June-October), in Maharashtra and UP during *rabi* (November-December) and in West UP and Central UP during spring (March-June). Since it is photoinsensitive and short duration crop it is an excellent catch crop to compensate the failure of crop the main.

Fig. 1 : Sunflower plant

BOTANY

Sunflower can grow very tall upto 15 feet depending upon the variety and season. However, presently we grow dwarf varieties that are only two or three feet long. The sunflower has a well developed tap root system which grows rapidly and penetrates nearly vertically downward (Fig. 2). Usually, its depth exceeds the height attained by the plant, and this holds true for all stages of growth. Numerous strong laterals appear early in the development of the plant. These mostly originate from the enlarged portion of taproot which usually occupies 4 to 6 inches of surface soil. Often, 30 to 45 large laterals occur on a single plant. These spread widely, usually 2 to 5 feet, some running in so nearly a horizontal direction. Others turn downward but seldom reach depths greater than 3 to 4 feet. The taproot, however, clothed throughout with small branchlets, reaches depths of 5 to 9 feet. The very numerous but mostly small branches clothing the laterals furnish an extensive and efficient absorbing system, especially in the surface 2 feet of soil. The roots are quite as much modified as are the tops by thick or thin planting. The cultivated type sunflower has single shoot, erect, thick grooved and hairy. The leaves are green, oval, hairy, serrated or notched margine covered with short-fairly stiff hairs, large, ranging from 20-40 in number depending upon the variety and season. The inflorescence is a capitulum or head, characteristic of the family, Asteraceae. Sunflower in bud stage exhibit heliotropism. At sunrise the faces of most sunflower are turned towards the east. The orientation of the sunflower to the sun is a direct result of the direction of light. The actual movement of the flower occurs as a result of motor cells in a structure called the pulvinus. The pulvinus is a region that is located where the flower head is connected to the stem. The motor cells get bigger or smaller depending on how much water they absorb or lose. Potassium is involved in all this. If the motor cell has a lot of potassium more water will go into the cell and the cell will inflate. With less potassium, the cell loses water and deflates. Which cells inflate and deflate causes the flower head to move as the sun moves. At night when the sun sets the flower head will move and face the direction the sun should rise in the next morning. This happens because of the amount of water contained in the motor cells of the pulvinus. The wild sunflower typically does not turn towards the sun. Its flowering heads may face many directions when mature.

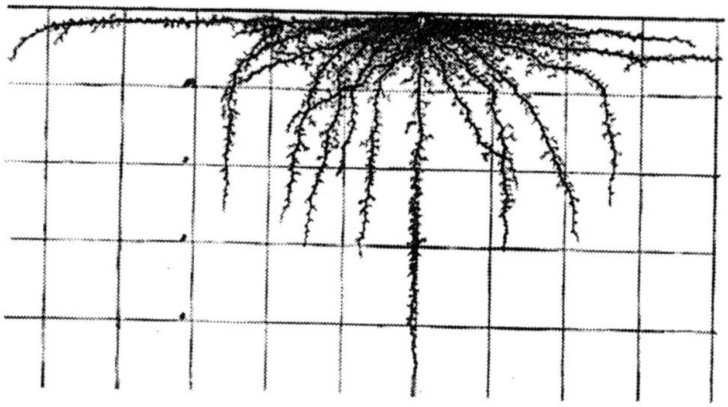

Fig. 2 : Root system of a 2.5-months-old sunflower

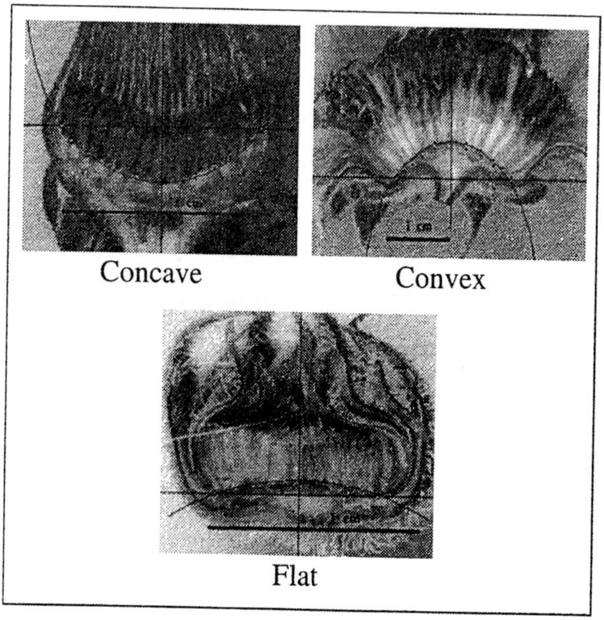

Fig. 3 : Types of sunflower head.

However, the leaves typically exhibit some heliotropism. Once the flower opens into radiance of yellow petals, it faces east. No one knows why. The sunflower stalk has a main or primary flower head and may have one to several secondary heads. All commercial cultivars are composed of almost 100 percent single headed plants. The main head may have from 1000 to 4000 individual florets depending upon the variety and size of head. Secondary heads may have Concave from 500 to 1500 florets. The sunflower head is of three types *i.e.* concave, convex and flat or plane (Fig. 3). Concave and flat shaped flower heads of medium size yield seeds with more oil (>45%), insaturation (IV >120) and linoleic acid (>48%) than did the concave shaped flower heads of medium size; 39.3% oil, (IV 112.6) unsaturation, 39.3% Linoleic acid, respectively (Table 1) (Afzalpurkar and Laksminarayana, 1980). Some of the heads face sun (sun loving), some turn opposite of sun (sun hating) and some are not responsive either negative or positive, remain neutral. This behaviour is not the characteristic of any particular variety.

Table 1 : Variations in Oil Content and Fatty Acid Compositing with Sunflower Head Size and Shape

			Fatty Acid Composition (Wt. %)				
Sunflower Size and Shape	*Oil[a] %*	*I.V. (calc.)*	*16:0*	*18:0*	*18:1*	*18:2*	*18:2÷ 187%1*
Size[b]							
Small (4" dia.)	44.5	121.0	7.0	4.2	37.4	51.3	1.37
Medium (8" dia.)	45.2	120.6	6.1	2.4	43.4	48.1	1.11
Big (12" dia.)	42.0	100.7	6.0	4.6	62.1	27.3	0.44
Shape[c]							
Convex	45.2	120.6	6.1	2.4	43.4	48.1	1.11
Flat	45.0	124.3	6.0	2.5	38.9	52.5	1.35
Concave	39.9	112.6	5.2	3.7	51.8	39.3	0.76

[a]Moisture-free basis; [b]Convex shape; [c]Medium size (8" dia.)

***Source* :** Afzalpurkar and Laksminarayana,1980.

The head consists of two types of flowers (florets) *i.e.* ray florets and disc florets. Ray florets form the outer whorl of the capitulum and are asexual (Fig. 4). They have vestigial styles and stigma without anthers and hence sterile. Disc florets are arranged in concentric circle of the receptacle and are hermaphrodite with an inferior ovary. The sepals are modified into two pappus scales. The five petals are united except at the tip to form a corolla tube. Stamens are free and are attached to the base of the corolla tube. The five anthers are fudged to form anther tube. The flower starts opening from the outer whorl of the capitulum and proceeds towards the centre of the capitulum (Fig. 5). The whole opening of flowers takes 5 to 10 days, depending upon the size and the season. Anthesis takes place between 5 and 8 AM. The pollen and nectar are quite attractive to bees and are viable for 12 hours. The stigma becomes receptive the following day and continues to be receptive for 2-3 days. The fruit is achene, 8-25 mm long, 4-13 mm broad and consists of a seed (kernel) and the outer pericarp (hull). The seed consists of seed coat, endosperm and embryo. Most embryo is cotyledons. The endosperm consists of single layer of aleurone cells. The colour of seeds is white, black, white pale, gray and brown white stripes. The seed contains oil ranging from 35-45%, protein 18-23%, carbohydrate 18% and ash 3-9%. Seed is rich is vitamin A, D and E, antioxidants and polyunsaturates (especially linoleic acid). The seed is highly nutritive and palatable. Rodole (1968) has described the significance of sunflower seed and has suggested that eating sunflower seed (a small handful) a day will help to discard the use of sunglasses and will also help to withstand the glare of tropical sun. He argued that sunflower seed has something else besides being rich in vitamins which exerts this mysterious power over the eyes. Sunflower is heliotropic and its head turns to west to absorb the last rays of dying sun and the seeds therefore are drenched with sun vitality. It has also been stated that various kinds of eye problems could be overcome by consumption of soaked sunflower raw seeds for a period of one month.

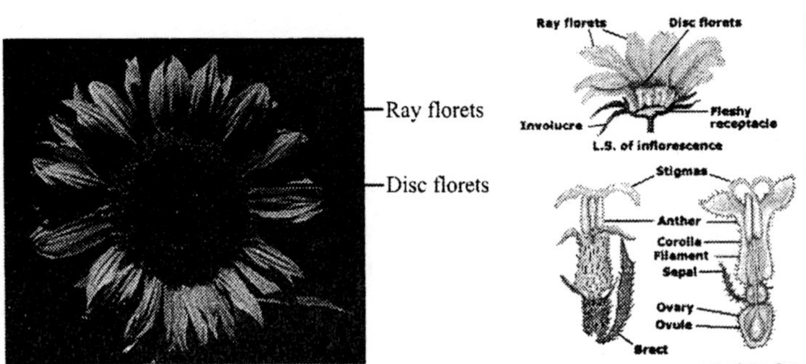

Fig. 4 : Sunflower head with LS of inflorescence and dist florets.

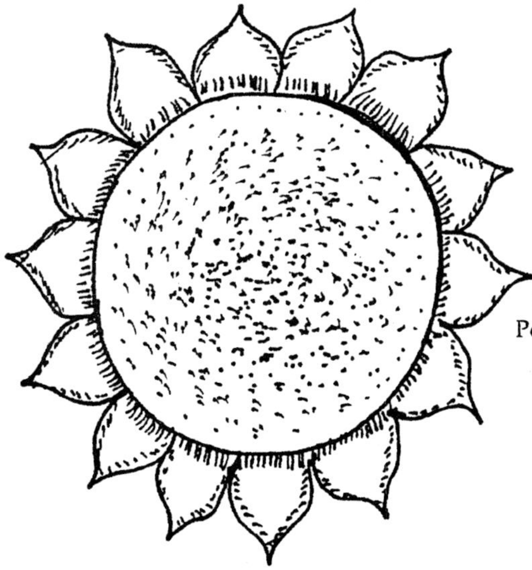

Fig. 5 : Flower opening in capitulum.

GROWTH AND DEVELOPMENT

Sunflower is a highly adoptable crop although this basically belongs to a temperate zone. It has been considered as indifferent to photoperiod (Allard and Garner, 1940) although varieties respond to photoperiodic treatments. Goyne *et al.* (1977), suggested it either indifferent or short day plant. In contrast, Anderson *et al.* (1978) reported variety Peredovik as developing faster in the vegetative stage under long days suggesting a long day responsive plant. In both cases, the data derived from field planting where photoperiod and temperature are strongly correlated and therefore do not provide of a true evidence of a true photoperiodic reaction. It grows and tolerates a temperature range of 8° to 34°C without significant reductions in yield indicating its wide adoptability to temperature.

Sunflower germination and establishment of seedlings takes approximately 15 days followed by leaf development which takes another 20 days followed by reproductive development. The flower bud initiation takes from 35 to 65 days and flowering from 65 to 80 days depending upon the varieties and season. Seed development takes 80 to 120 days (Fig. 6).

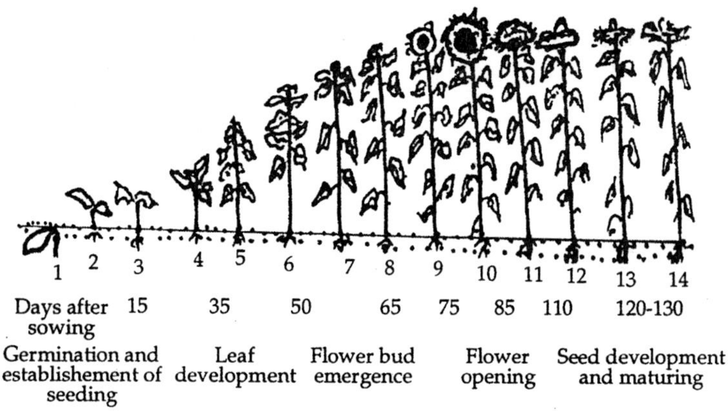

Fig. 6 : Ontogenic development in sunflower.

Sunflower is a short duration crop of approximately (100-120 days) from the date of sowing depending on the variety and temperature. It may take upto 40 days to reach an LA1 of 1 (Connor *et al.,* 1985) and an associated 80% radiation interception (Rawson *et al.,* 1984), even at high temperature with irrigation. Thus much of the crop life solar radiation is being utilized in biomass production. Net assimilation rates of sunflower plants grown widely spaced with soil nutrients and water non limiting, reached 2.0 g dm^{-3} wk^{-1} in clear weather at mid summer in an arid climates (Warren-Wilson, 1966). This suggests that maximum rates of photosynthesis in the most active leaves were 50-65 mg CO_2 dm^{-2} h^{-1}. These high rates are a response to the high level of radiation in the arid climate. They imply that (under non limiting soil) plants can attain higher productivity in arid climate than in any other.

Sunflower has very rapid growth and this is associated with higher photosynthetic rate and low carbon isotope discrimination (or high transpiration efficiency). Studies showed that carbon isotope discrimination (") was negatively * related to transportation efficiency and biomass accumulation amongst sunflower (*Helianthus annuus* L.) genotypes (Virgona *et al.*, 1990). While studying on a large number of genotypes of sunflower, Virgona and Farquhar (1996) suggested that if breeding programme for more transpiration efficient sunflower cultivars based on low isotope discrimination were established, the result would be a faster growing plant with a higher photosynthetic rate. However, it is not yet possible to establish whether faster growing plant would also be a higher yielding plant?

NUTRIENT USE EFFICIENCY

The supply of adequate nutrient to sunflower crop is of high importance. However, excess or insufficient nutrient availability adversely affect seed yield. The nutrient use efficiency is defined as seed yielded per unit of nutrient supply. For nitrogen use efficiency, it is the seed yielded per unit of nitrogen supply. In sunflower yield and nitrogen use efficiency depend on N uptake from the soil and its remobilization of stored vegetative nitrogen accumulated before the flowering stage. This nitrogen remobilization can escalate, particularly under hot and dry weather after anthesis (Escalante *et al.*, 1998), but sunflower plants can also absorb a high amount of nitrogen from the soil, resulting in high nitrogen use efficiency and decreasing the risk of nitrate loss into the ground water (Baldini *et al.*, 1996). Therefore, nitrogen supply in the first vegetative stages of sunflower growth could play an important role in yield and N utilization. The growth and development of sunflower is better with nitrate as source of nitrogen (Kurvits and Kirby, 2007). The uptake of K, Ca, Mg and Na per unit dry weight were also higher in nitrate fed plants. The requirement for nitrogen, phosphorus and potash, the major nutrients for sunflower as external application depends on the nutrient status of soil, temperature and water availability. There are different reports regarding the requirement of nitrogen, phosphorus and potash. For example, Bakht *et al.* (2010) have reported maximum leaf area index (LAI), crop growth rate (CGR), dry matter accumulation (DMA) and seed yield with application of 150:100:100kg ha^{-1} nitrogen, phosphorus and potash, respectively. Contrarily Montemurrod and Giorgio (2005) have reported that application of 50 kg N ha^{-1} and 100 kg N ha^{-1} has similar effect on most of the parameters and had no effect on seed yield. While Thavaprakash *et al.* (2003) have reported the highest yield of sunflower with N/P ratio of 1.0 in the treatment receiving 120 kg N and 120 kg P_2O_5 ha^{-1}. High nitrogen content however not only increases the vegetative surface, but also pollutes the ground level water. A balance nitrogen and other nutrients supply need to be worked out so that maximum yield with minimum nutrients input could be obtained. Phosphorus in soil was found to vary significantly only in low P status. Critical phosphorus level was reported to be 20 kg P ha^{-1} below which sunflower may respond to phosphorus level (Muralidharudu *et al.* 2003). Sulfur(s) forms important partnership with N in sunflower production. Together they determine leaf area, which provide photosynthates to developing florets and seed, following through to yield and seed size. Deficiency when combined with N reduces seed yield and number of seed. The amount of sulfur removed in the seed of a one tonne crop is 5 kg ha^{-1} with uptake highest between budding and anthesis, a large proportion is also taken up post anthesis (Anonymous, 2008) (Table 2).

Table 2 : Sulphur Uptake by Sunflower

Growth stage	*% uptake*
Emergence-budding	20
Budding to anthesis	45
Post anthesis	35

Source : Anonymous (2008)

Deficiency of sulphur delays the floret initiation and anthesis but not seed maturity (Hocking *et al.*, 1987). Potassium is required for stalk and tissue strength. Sunflowers are a high user of K with every ton/ha yield removing 30 kg/ha.

WATER REQUIREMENT

Water requirement of sunflower compared to many cereals is high and vary from 600 to 1000 kg of water to produce dry matter/kg depending on the climate and total growing period. Evapotranspiration increases from establishment of flowering and can be as high as 12 to 15 mm/day. High evapotranspiration rates are maintained during seed setting and ripening period. Total crop water use over different growth period has been reported to be 20 per cent during vegetative period, 55 per cent during flowering and remaining 25 per cent during yield formation and ripening period. Crop coefficient (Ke) is simply ratio of evapotranspiration (ET) observed for the crop studied over that for the reference crop under same condition. The crop coefficient is 0.3-0.4 during initial stages (20-25 days), 0.7-0.8 during crop establishment stage (35-40 days), 1.05-1.2 during the mid season stage (40-50 days), 0.7-0.8 during the late season stage (25-30 days) and 0.4 at maturity to harvest.

Although water requirements of sunflower is relatively high compared to most other crops, the crop has an ability to withstand short period of severe soil moisture deficit of upto 15 atm tension. Long period of severe water deficit at any growth period cause leaf drying with subsequent reduction in seed yield.

ABIOTIC STRESS

Sunflower is considered a fairly good drought resistant crop. However, severe water deficit during early vegetative period results in reduced plant height but may increase root length. Adequate water during late vegetative period is required for proper bud development. The flowering period is the most sensitive to water deficit which cause considerable yield decrease since fewer flowers only come to full development. Yield formation is equally sensitive period to water deficit, causing severe reduction in yield and oil content.

Various morphological and physiological processes and yield characters are affected severely as a result of water stress (Deshmukh *et al.*, 1988). Significant reduction in the total chlorophyll content and photosynthesis in the leaves of sunflower has been reported at a leaf water potential of -15 to -25 MPa (Manamohan, 1995). Nitrate reductase (NR) enzyme has been reduced in the leaves of sunflower as a result of water stress at any growth stage, however the extent of reduction varies depending upon the growth stage (Deshmukh *et al.*, 1988). It indicates the significance and sensitivity of an important step in nitrogen metabolism, a most important process which is vulnerable to water stress. The reduction in photosynthesis was due to both; higher stomatal resistance and reduction in photosynthetic process (Table 3 to 6).

The stomatal factors were observed to be more susceptible at vegetative stage whereas non stomatal factors becomes more sensitive at flowering stage (Table 6).

Manmohan (1995) has reported that in sunflower there is reduction in Rubisco protein due to water stress (Table 7).

Among several traits being followed for evaluation of plant water status, relative water content (RWC) is the best measure of the level of water deficit in the plant at a specific point of time. RWC is related to cell volume and closely reflects the balance between water supply to the leaf and transpiration rate (Sinclair and Lidlow, 1985). Other traits such as leaf turgor, osmotic potential are related to osmotic adjustment (Perniola, 1994). For osmotic adjustment, plants accumulate certain type of compatible solutes under stress environment to lower its potential. As a result of lower potential, plant, may regain its turgor and becomes capable to attract more water from the soil. The value of osmotic adjustment (OA) as a trait in sunflower which can contribute effectively to yield maintenance under drought has been reported by Cheminti *et al.* (2002). Working with a number of inbred lines differing in drought tolerance, they have concluded that OA can contribute to drought tolerance in sunflower and stressed the need to develop markers for this traits. Compatible solutes help in maintaining cell turgor, protect against denaturating important structural molecules and cell organelles and the concentration does not cause toxicity in cell. There are many molecules which may be considered as compatible solutes such as K^+, Ca^{++}, proline HSPs and dehydrins. However, in sunflower, several reports suggest the accumulation of free proline which is an important osmolyte under drought stress and also because of its structure which

Table 3 : Effect of Moisture Stress on Stomatal Conductance* for CO_2 (cm m^{-1}) and its Subsequent Recovery, 72 hr after Rewatering, at Various Stages of Growth and Development in Sunflower (Var. Morden)

ψ_1 (bar)	*Vegetative*				*Flowering*			
	Stress	*% of control*	*Re-watered*	*% of control*	*Stress*	*% of control*	*Re-watered*	*% of control*
-6 ± 1	0.638(0.256)	100	0.976(0.391)	100	0.336(0.135)	100	0.406(0.163)	100
-12 ± 1	0.129(0.052)	20	0.512(0.205)	52	0.152(0.061)	45	0.316(0.127)	78
-18 ± 1	0.057(0.023)	9	0.313(0.125)	32	0.062(0.025)	18	0.164(0.066)	40
-24 ± 1	0.036(0.014)	6	0.155(0.062)	16	0.031(0.012)	9	0.126(0.050)	31

Figures within parenthesis indicate the corresponding values in mol m^{-2} s^{-1}

*Conversion factor for converting the stomatal conductance from cm s^{-1} to mol m^{-2} s^{-1} was taken to be 0.4007 (Farquhar *et al.*, 1978; Wong *et al.*, 1979)

Source : Prasad (1990), Prasad and Srivastava (1990)

Table 4 : Effect of Moisture Stress on Net Photosynthesis (μmol CO_2 m^{-2} s^{-1}) and its Subsequent Recovery, 72 hr After Rewatering, at Various Stages of Growth and Development in Sunflower (Var. Morden)

ψ_1 (bar)	*Vegetative*				*Flowering*			
	Stress	*% of control*	*Re-watered*	*% of control*	*Stress*	*% of control*	*Re-watered*	*% of control*
-6 ± 1	20.72	100	20.62	100	20.81	100	20.34	100
-12 ± 1	15.42	74	17.99	87	14.51	70	15.78	77
-18 ± 1	11.14	54	15.90	77	9.26	44	11.24	55
-24 ± 1	4.75	23	14.29	69	2.54	12	3.91	18

Source : Prasad (1990), Prasad and Srivastava (1990)

Table 5 : Effect of Moisture Stress on PSII (μequ e-hr^{-1} mg chl^{-1}) and its Subsequent Recovery, 72 hr After Rewatering, at Various Stages of Growth and Development in Sunflower (Var. Morden)

ψ_1 (bar)	*Vegetative*				*Flowering*			
	Stress	*% of control*	*Re-watered*	*% of control*	*Stress*	*% of control*	*Re-watered*	*% of control*
-6 ± 1	356	100	356	100	450	100	438	100
-12 ± 1	350	98	360	100	330	73	438	100
-18 ± 1	298	84	282	79	180	40	285	65
-24 ± 1	238	67	198	56	140	31	132	30

Source : Prasad (1990), Prasad and Srivastava (1990).

Table 6 : Effect of Moisture Stress on Stomatal and Non-Stomatal Limitations to Photosynthesis at Various Stages of Growth and Development in Sunflower (Var. Morden)

ψ_1 (bar)	*Per cent of total inhibition to photosynthesis caused by*			
	Stomatal Component	*Non-stomatal Component*	*Stomatal Component*	*Non-stomatal Non-stomatal*
	Vegetative		*Flowering*	
-6 ± 1	9	91	10	90
-12 ± 1	35	65	28	72
-18 ± 1	57	43	44	56
-24 ± 1	38	62	24	76

Source : Prasad (1990), Prasad and Srivastava (1990).

Table 7 : Effect of Water Stress on the Amount of Rubisco Protein in Two Sunflower Cultivar

Treatment	*Morden* *RWC*	*Rubisco Protein (%)*	*Treatment*	*MSFH-8* *RWC*	*Rubisco Protein (%)*
Control	–	100.00	Control	–	100
Stress (Low)	83.78	89.47	Stress (Low)	85.75	91.52
-10 bars	(±1.56)	(±1.79)	–10 bars	(±0.84)	(±2.42)
Stress (Medium)	74.56	86.94	Stress (Medium)	71.78	86.51
-15 bars	(±1.02)	(±2.23)	–15 bars	(±1.27)	(±2.19)
Stress (Severe)	61.43	79.70	Stress (Severe)	59.22	82.23
-25 bars	(±0.97)	(±1.85)	–25 bars	(±1.43)	(±1.96)

Values in parentheses are ±SEm

Source: Manmohan (1995)

increases water retention capacity of the cell (Gasev *et al.*, 1968). However, there are conflicting results on the involvement of proline in drought tolerance. Deshmukh *et al.*, (1988) have reported that proline accumulation did not correlate with the stress tolerance in sunflower, it rather indicates an index of stress level. A screening for drought resistance has been suggested on the basis of NR enzyme (Deshmukh and Srivastava 1983, Deshmukh *et al.*, 1988) and stay green characters (Skoric, 2009)

Along with drought, salinity is also important environmental stress that alters the plant water status and severely affect crop productivity. The accumulation of compatible solutes is one of the strategies that plants have developed to tolerate salt stress. Proline and betaine are the main metabolites that accumulate in various species of higher plants in response to salt stress. In *Helianthus annuus* L., pre-treatment of seeds with ethanolamine led to enhanced seedling tolerance to conditions of saline stress during germination, as evidenced by the greater growth of pretreated seedling. During the germination period, a considerable increase was observed in proline levels (upto 300%) in seedlings subjected to saline stress (Kogan *et al.* 2000).

In the recent past a number of drought responsive (Ingram and Birtels, 1996, Kim *et al.*, 2000, Nepomuceno *et al.*, 2000) and salinity responsive (Moons *et al.*, 1997, Wei *et al.*, 2000) genes were cloned and characterized from different plant species. Transcription of many of these genes e.g. LEA proteins (Espelund *et al.*, 1995), glyceraldehyde-3-phosphate dehydrogenase (Jeong *et al.*, 2000), ABA responsive element binding proteins AREB-1 and AREB-2 (Uno *et al.*, 2000) are upregulated by both drought and salinity. Expression of other genes is regulated specifically by either drought or high salinity.

Many investigations for screening aspects on a range of crop' development under *in vitro* and *in vivo* conditions have been carried out to determine the responses to salinity or drought stress (Lynch and Tai, 1989, Haverkkort *et al.*, 1990, Levy, 1992). Screening plays a crucial role in breeding programmes for inducing salinity tolerance. According to a classification based on water stress day index. Sunflower was determined as a moderately sensitive crop to salinity (Katarji *et al.*, 2003). Francois (1996) has called it a moderately tolerant crop. The concept of water stress day index (WSDI) provides a quantitative method for determining the stress imposed on a crop during its growing season (Hiler and Clark, 1971). Oil concentration and seed was relatively unaffected by increased soil salinity upto 10.2 dSm^{-1} for growth under moderately saline soil condition. There has been variation in response of sunflower genotypes to salinity (Ashraf & Tufail, 1995, Muralidharudu *et al.*, 1999). For economical production in saline soils, it is important that many commercially released sunflower cultivars require to be tested for salinity using a rapid reliable screening method. As a result, in terms of either development of salt tolerant plants or determination of suitable salt tolerant crops for a region, selection and evaluation of salt tolerance of plants has a preliminary importance. Turhan and Ayaz (2004) have suggested the screening for tolerance at seedling emergence stages and growth of sunflower which can be a practical approach for screening for reducing plant material prior to field experiments.

The role of potassium has been shown to be beneficial under salt condition. In an experiment, sunflower seeds were grown only with increasing saline solution 10, 50 and 100 mM sodium chloride (NaCl) and potassium (K) supply to determine how salinity and potassium affect germination and growth. Potassium supply under saline condition (100 mM) or non saline condition had a beneficial effect on sunflower seedling germination which was not significantly altered with moderate salt concentration (50 mM). During this stage, K supply in absence of salinity increased significantly seedling biomass which reflects what is happening in the aerial part and root. In a saline environments K supply did not markedly alter plant dry matter production.

Higher than normal rainfall during grain filling of sunflower has been shown to be associated with reduction in grain filling because of low radiation (cloudy weather) or also due to increased incidence of diseases. The leaf area is reduced and it also reduces the photosynthesis of the remaining leaves (Grassini *et al.*, 2007). The effect of water logging is not uniform it experienced at different growth stages. For example Orchard and Jessop (2005) have reported that water logging at 6 leaf stage and bud visible stages, the leaf expansion and stem extension are inhibited, although these effects did not always persist until maturity. While with anthesis, water logging causes rapid desiccation of leaves. Yield was most affected by the anthesis water logging but no consistent effect on seed number or 1000 seed weight.

Our concept of DNA based markers has revolutionized our ability to directly access any part of the plant genome and has led to new opportunity such as screening a large number of segregating population for QTL (Quantitative trait loci) related to tolerance without exposing them to stress and independence of plant growth stage. In nature the vast majority of physiological traits are QTL. QTL show continuous phenotypes distribution affected by a large number of loci and show significant interaction with environment (Tanksley, 1993). In sunflower however very few studies have been made to determine the development of a molecular marker for QTL related to abiotic stress tolerance (Kiani *et al.*, 2007b).

PHOTOSYNTHESIS

Sunflower although being a C_3 plant is highly efficient in photosynthesis. Warren Wilson (1966) reported net assimilation rates of 2.0 g dm^{-3} wk^{-1} in sunflower grown widely spaced with soil nutrient and water no limiting condition in clear weather at mid summer in arid climate. This implies that under non limiting soil, plants can attain higher productivity in the arid climate than in any other. Genetic variation in rate of photosynthesis in sunflower has been reported for cultivars and possible causes of difference has been suggested as protein, pigments and stomatal frequency. Photosynthesis in leaves differs due to their position and time of the day (Bana and Srivastava, 1987). Prasad *et al.* (1989) have reported that the highest photosynthetic activity is in the first fully developed leaf followed by second and third leaf. The rate of photosynthesis in all the leaves was high during the forenoon and declined in the afternoon (Fig. 7). It was further noticed that the older leaves (*i.e.* L_1 and L_2) which were very close to the base of the plant, had comparatively low photosynthetic rate and maintained it only for a short period (upto 12.00 noon), whereas younger leaves which were on the top of the plant showed relatively high photosynthetic activity for a longer period (upto 4.30 pm). Diffusive resistance although did not show any definite trend with photosynthetic rate, however in general, the older leaves recorded comparatively high leaf diffusive resistance. English *et al.* (1979) have reported the maximum photosynthesis in the larger fully expanded leaves. These fully developed leaves are responsible for supplying most of the assimilate required by the developing seeds (McWilliam *et al.*, 1974). Older leaves at the bottom of the canopy and small immature leaf at the top of the stem had lower rate of photosynthesis and this could partly be due to lower leaf conductance. (Goswami and Srivastava, 1988; Srivastava and Goswami, 1988). All the leaves have increasing rate of photosynthesis after emergence to a maximum and then declines prior to senescence.

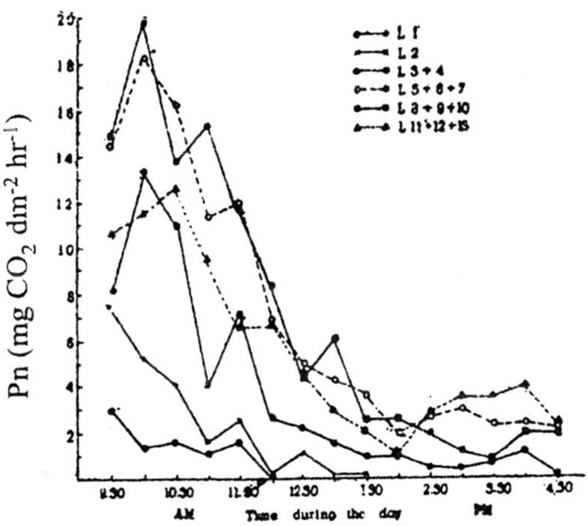

Fig. 7 : Net photosynthetic rate of different leaves at different times of the day in sunflower.

Source : Prasad *et al.* (1989).

Photosynthesis has a diurnal change and is limited under many circumstances by its product. It the rates of export are not sufficient to prevent assimilate accumulation in source leaves, the photosynthesis rate will decline. The synthesis of sucrose (instrumental in regulating photosynthesis) in leaves involves cytoplasmic enzyme sucrose phosphate synthase (SPS), the inhibition of which would lead to depletion in the availability of inorganic phosphate (Pi) in the cytosol, thereby affecting the movement of triose phosphates from the chloroplast to the cytoplasm and thus sucrose synthesis is affected. SPS thus seems to play a pivotal role in regulating the photosynthesis. However, in sunflower Mitra and Srivastava (1993 and 1994) have reported that SPS does not correlate with photosynthesis (Table 7, Fig. 8).

Table 7 : Diurnal Variation in Net Photosynthetic Rate (Pn), Sucrose Phosphate Synthase (SPS) and Sucrose Content in Sunflower. Mean of Three Replication ±S.D.

Time	*Pn [mg(CO_2) $m^{-2}s^{-1}$]*	*SPS [μmol (sucrose) kg^{-1} (f.m.) s^{-1}]*	*Sucrose [g kg^{-1} (f.m.)*
06.30	0*	10.9 + 1.4	1.83 + 0.17
12.00	0.548 + 0.036	19.0 + 4.4	4.93 + 0.25
16.00	0.243 + 0.025	19.5 + 1.4	4.65 + 0.33
20.30	0*	14.0 + 1.5	3.18 + 0.13

* not measurable.

Source : Mitra and Srivastava (1993).

It appears that SPS in sunflower is a constitutive enzyme and is not affected by day and night.

Sunflower can grow in a wide range of temperature but optimum photosynthesis is from 27 and 35°C (English *et al.* 1979). The photosynthesis in sunflower leaves decreases below 20°C and above 40°C significantly. Photosynthesis in sunflower increases when measured from low to high intensity (English *et al.* 1979).

Despite the high photosynthetic capacity of sunflower, the crop is not particularly efficient in partitioning of carbon assimilates into the developing seed. The harvest index is extremely poor of around 33% or even less in Indian condition which is lower that most values reported for modern cereal varieties (Firscher and Kertesz, 1976). The most effective way to improve partiting is to reduce the shoot weight and increase the number and weight of seed per inflorescence.

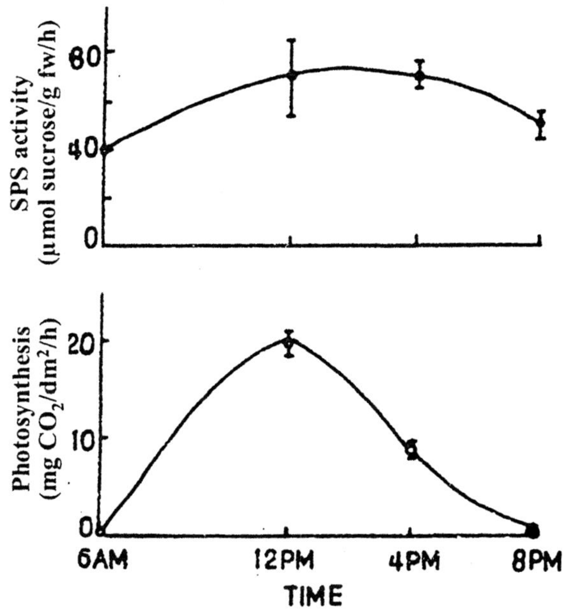

Fig. 8 : Diurnal changes in photosynthesis and SPS activity in sunflower Mean of three replication ±S.D.

SEED GROWTH AND OIL SYNTHESIS

In sunflower (*Helianthus annuus* L.), the position of the capitulum during grain-filling differs among plants even in same cultivar. Some plant maintain the head in a vertical position during the whole of the grain filling period, while others hide their heads within the canopy after anthesis. These differences in head position influence thermal regime, warmer regimes shortening the grain filling period and reducing grain dry weight (Ploschuk and Hall, 1995).

Sunflower grain is comprised of pericarp and kernel (embryo) (Connor and Hall, 1997). The pericarp comprises about

20% of final grain weight, so the effects of environmental factors on embryo growth have greater implications for seed yield and quality.

Seed development in sunflower can be divided in three sequential phases (Lindstrom *et al.,* 2002). The initial phase (phase I) upto 10 days after anthesis includes fertilization, the period of active cell division and a slow increase of dry weight. Also during this phase the pericarp completes its development (Connor and Hall, 1997). Next follows an intermediate phase (phase II) where there is an increase in cell volume and a rapid accumulation of dry matter in seed. In the final phase (phase III) maximum seed weight is achieved and the fruit reaches physiological maturity.

In sunflower there is a developmental and growth gradient which progresses from the peripheral seed towards those at the centre. Therefore, any effect either environmental or nutritional while the reproductive development is in progress will not be similar for fruits at different position in the capilulum. For example, Yegappan *et al.* (1982) observed that water stress in post anthesis reduced the seed weight in the central portion and not in the fruits at the peripheral region. Assimilate supply if reduced due to senescing leaves during seed filling period will also affect the seed filling in the centrally located seeds (Goswami and Srivastava, 1987).

Therefore, the assimilate requirements of all the seeds in the capitulum must be balanced with the actual assimilate supply by the leaves. Selection of long seed filling period with adequate assimilate supply will increase the seed yield.

The synthesis of oil in sunflower seed starts following fertilization however the oil synthesis is slow initially but follows a rapid increase from 10 to 25 days and thereafter declines. There are different reports about the period of synthesis of oil in sunflower which varies in different cultivars. Reserve lipids in sunflower and other oilseeds are laid down in sub cellular structures known as oil bodies formed by a hydrophobic matrix of triglyarols surrounded by a larger of phospholipids, on which proteins called (Oleosins and Caleosins) are deposited (Huang,

1992, Murphy, 2001). Oleosins and the phospholipids layer confer stability to the oil bodies and maintain them as separate entities. In one study in sunflower, Beandoin *et al.* (1999) revealed that oleosins are synthesized on endoplasmic reticulum bound ribosomes and later transformed to areas of the endoplasmic reticulum involved in the oil body biosynthesis.

Besides oil, the embryo cells of sunflower contain protein bodies as a reserve component (Bolyakina and Raikhman, 1999). Ploschuk and Hall (1997) reported that protein accumulation in sunflower occurs at a greater rate during the first part of the achene filling period. Achene oil and protein concentration in the mature achene are often found to exhibit a negative association (Connor and Sadras, 1992, Lopez Pereira *et al.*, 2000).

The synthesis of oil in seed starts almost immediately after fertilization and oil percentage increases to maximum over the next 20-25 days, several weeks before the seed reaches its final dry weight.

PHYSIOLOGICAL ANALYSIS OF LOW SEED YIELD

Fig. 9 : Location of filled and unfilled seeds in capitulum.

Inspite of the fact that sunflower is a good short duration crop and its oil is nutritionally rich, its cultivation in India has not received proper attention due to poor yield. The poor seed yield is due to poor seed filling.

The yield of sunflower as compared to cereal is low although it is not expected to compete with cereals. Abelardo *et al.* (2002) has reported the yield of 2204-2399 kg ha^{-1} in Argentina. In India, the yield of old varieties is 600 to 1000 kg ha^{-1} and of some of the hybrids is 1000 to 2700 kg ha^{-1}. However, this yield is never achieved in the farmers field. The yield in temperate and subtropical region is normally higher due to the fact that the crop duration is longer than in tropical region. The oil content in sunflower is comparatively much higher than many other oilseed crop besides being safe and nutritionally rich. In India the oil content varies from 38 to 46% depending upon the variety and season in which it is grown.

A large number of sunflower seeds remains unfilled or partially filled and thus, seed filling is an important problem in cultivation of sunflower. A large portion particularly in the central region of the capitulum remains unfilled (Fig. 9) or partially filled reducing the total seed yield per plant per hectare. Several reasons have been attributed to the seed filling problem.

Pollination

One of the reason has been attributed to lack of pollination (Free, 1964). Sunflower is highly cross pollinated plant and pollination is carried out by bees. Since sunflower pollen is heavy and sticky so cannot be carried by wind. Both pollen and vector are quite attractive to bees throughout the day and are very favourable for cross pollination. However, pollination can be greatly affected by the high temperature and frequent shower during flowering, since the activity of bees may be reduced or the pollen germination may also get reduced along with pollen breakdown (Free, 1964).

The opening of all the florets in a single head takes 5 to 10 days or even more (Fig. 9). Pollination must occur during this period otherwise there will be delay in fertilization. Sanduleac (1960) compared seed production from four pollination treatments (1) Isolated under gauze (2) Isolated under gauze and hand pollination four times (3) Open to insect pollination and (4) Open to insect pollination and hand pollination four times. Seed production was always greatest on open plots, with no increase obtained from the extra hand pollination Avetisyan (1965) suggested that for the best seed production, each floret should be visited 8 to 10 times. In Argentina Schelotto and Pereyras (1971) showed that seed yield increased five to six times and the oil content of the seed increased by 25 per cent in plots exposed to honey bee colonies as compared to the yield and oil content of the crop from the plot isolated from insects during flowering. The exact number of bees needed for maximum pollination has not been determined. Noetzel (1968) placed one colony of bees per acre at the end of different sunflower fields in North Dakota and counted the bees per head and obtained seed production data at different\distances from the apiary. He obtained overall increase of 20 per cent due to the bees alone but beyond 400 feet from the apiary, he got no measurable increase in yield.

Studies on sunflower pollination, indicate that honey bees are the pollinating agent, and that colonies should be provided to the field (Barbier and Abid, 1966) and also, that they should be protected from harmful pesticides while they are in the field. If adequate pollination throughout the field provided, there should be no significant gradient of seed set in relation to apiary location.

Self Incompatability

A number of workers believe that one of the major factors for hollowness in seed of sunflower is the self incompatability and insufficient pollination. If pollen from a floret is transferred to the receptive stigma of another flower on the same head and the flowers are self compatable, seed will set. However, most cultivars are self incompatable (Free and Simpson, 1964) in which case the pollen must come from another plant. When selfing (within the capitulum) occurs, seed setting is normally low (Barret 1954, Cardon 1922 and Kalton, 1951), the seeds are undersized and the oil content and germination are reduced. Selfed seed also sprout more slowly and production from them is lower than from plants derived from crossed seed.

ASSIMILATE LIMITATION/SOURCE-SINK RELATIONSHIP

Besides the insufficient pollination and incompatibility many workers believe that poor seed set setting and partially filled seed set in the central or middle portion of the capitulum is due to lack of sufficient assimilate available for seed filling. In sunflower, excessive vegetative growth occurs (Fig. 10) which is not conducive for better seed filling as a huge proportion of the photosynthate is diverted for the building of the vegetative surface even at the time when seed filling is in progress. Thus there seems to be a competition as regards assimilate for the growth of vegetative parts and for the growing seeds in the capitulum.

Fig. 10 : Plant with excessive vegetation.

In sunflower, photosynthates appears to be limiting in spite of the large green surface and high net assimilation rate relative to other species under full sunlight conditions. Estimations for CER indicate that leaf photosynthesis and Rubisco enzyme activity during the plant ontogeny increases from seedling to grain setting stage but declines before the seed filling is complete (Fig. 11). It is plausible that most of the assimilate is utilized for bullding of the vegetative surface resulting in limitation of the photosynthates for the developing seeds. Only the top five to three fully developed leaves and bracts in proximity to the capitulum contribute assimilates significantly towards the seed filling. The declining trend in photosynthesis during the seed filling period has been attributed to loss in ribulose *bis*phosphate carboxylase (Rubisco) enzyme activity and increase in protease activity (Table 8). Thus, it indicates that assimilate is limiting for seed filling in sunflower.

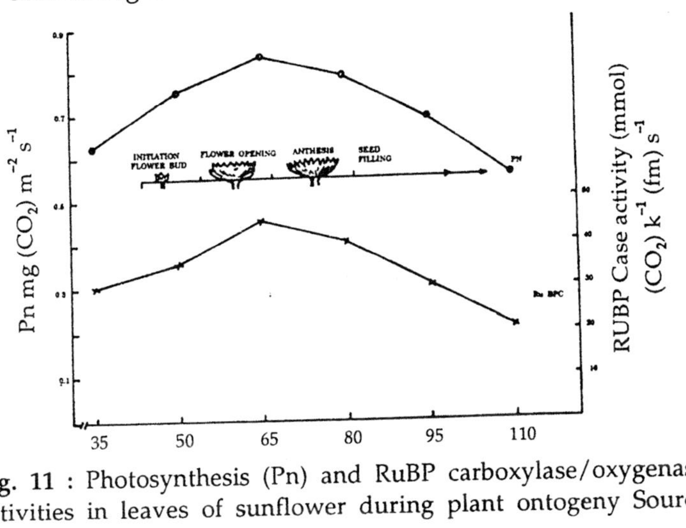

Fig. 11 : Photosynthesis (Pn) and RuBP carboxylase/oxygenase activities in leaves of sunflower during plant ontogeny Source Anonymous (1991).

Foliar spray of nitrogen, in the form of urea (0.05%) and benzyl adenine a synthetic growth regulator three times during the growing season delayed the loss in chlorophyll, RuBPcase activity and net photosynthetic carbon fixation (Goswami and Srivastava, 1988). This resulted in enhanced seed yield (Goswami and Srivastava 1987) (Table 9). Prolongation of post

Table 8 : Total Chlorophyll, Soluble Protein, RuBP Carboxylase *in vitro* and Protease Activity in the Leaves of Sunflower at Different Growth and Developmental Stages

	Day After Sowing					
	35	*50*	*65*	*80*	*95*	*110*
Total chlorophyll (mg/g fwt)	2.42	2.12	1.86	1.49	0.96	0.79
Soluble protein (mg/g fwt)	16.75±1.37	17.25±1.36	13.50±1.36	12.75±1.71	17.25±1.27	10.62±1.51
RuBP carboxylase (µmol CO_2/h/g fwt)	143.5±2.9	172.1±3.7	109.8±2.2	89.3±2.1	82.8±2.1	69.4±2.3
Nitrate reductase (µmol NO_2 formed/h/g fwt)	5.41±0.43	3.70±0.37	2.24±0.22	1.88±0.21	1.39±0.28	0.85±0.09
Protease activity (OD_{570}/g fwt/ 90 min)	11.2±1.2	14.5±1.8	16.21±1.4	16.45±1.8	17.49±1.4	18.6±1.2

Source: Goswami (1985)

Table 9 : Seed Weight (g) Per Plant at Harvest as Influenced by BA and Urea Treatments at Different Stages of Growth and Development; Average of 5 Replication in Each Case

Treatments	*Time of Application (Days After Sowing)*					
	30	*60*	*90*	*60+90*	*30+60+90*	*Mean*
BA (50 ppm)	11.87	12.25	11.57	13.23	13.97	12.56
BA (100 ppm)	11.65	13.13	10.80	14.34	14.89	12.96
Urea (0.5%)	11.23	12.38	10.67	12.90	13.13	12.06
Control	11.52	11.52	11.52	11.52	11.52	11.52
CD at 5% for :						
Control × other treatments	1.01					
BA concentrations	0.554					
BA × urea	0.480					
Time of application	1.23					
BA concentration x time of application	0.91					
Treatment × time of application	0.81					

Source : Goswami and Srivastava (1987)

flowering active photosynthesis is likely to enhance seed yield further and the possibility of extending it through genetic manipulation needs to be explored.

Most of the morphological characters show to bear a positive correlation with seed yield (Deshmukh and Srivastva, 1986). Stem weight particularly possessed positive correlation with seed yield and oil content under irrigated field condition. Chandra and Anand (1977) have observed that the area of unfilled seeds in the head of sunflower was negatively correlated with the thickness of stem. They have argued that seeds develop better when sufficient nutrients are supplied through a thick stem. While analyzing the seed filling in sunflower Khanna (1972) have reported a negative correlation between stem diameter and capitulum size and seed set. Russell (1953) had earlier reported that rise in plant height decreases seed yield. He has shown a significant regression of shoot height and stem diameter (b = 23.12 very high) and a positive correlation between stem height and leaf area (y = 0.7278) of the plant. This indicates that an increase in plant height results in increase in total leaf number and area results in increased vegetative surface which results in low photosynthetic activity due to mutual leaf shading. Besides it also restrict the supply of nutrients to seed. Very tall plants not only have more vegetative surface but also prone to lodging. However an optimum plant height must be there to build reasonable leaf surface for photosynthesis.

A correlation of morphological characters with seed yield has been worked out with 36 genotypes of sunflower grown under irrigated field conditions (Deshmukh *et al.* 1986) (Table 10).

Higher leaf surface has not been found to provide high seed yield, however, the significance of leaf and optimum leaf area can not be overlooked. Kioezowsk (1978) has reported the optimum LAI of 3.0 and Rawson *et al.* (1980) have shown a linear increase in sunflower seed yield between LAI of 0.1 and 3.0. However, Francis Flenet *et al.* (1996) have reported maximum LAI of 2.34 to 2.98 recorded at three growth stages. The optimum leaf area index is very important for longer maintenance of photosynthesis and thus for higher yield.

Table 10 : Correlation Coefficient Among Morphological Characters in Sunflower Genotypes Under Irrigated Condition

	Plant Height	*Stem Diameter*	*Head Weight*	*Head Weight*	*Seed Weight*	*100-Seed Weight*	*Seed-Husk Ratio*	*% Filled Grains*	*Stem Weight*	*% Oil*
Plant	1.0000									
Stem diameter	0.6519**	1.0000								
Head diameter	0.4414**	0.7737**	1.0000							
Head weight	0.4425**	0.7139**	0.8393**	1.0000						
Seed weight	0.4603**	0.6776**	0.7522**	0.8573**	1.0000					
100-seed weight	0.1977	0.3856**	0.5500**	0.6090**	0.4167**	1.0000				
Seed-husk ratio	–0.1111	-0.2095	–0.1047	-0.0252	–0.0259	0.0069	1.0000			
% filled grains	0.2468*	-0.2571**	0.1389	0.0916	0.2200*	0.5124**	0.1668	1.0000		
Stem weight	0.5124**	0.6272**	0.5703**	0.5039**	0.4581**	0.3548**	–0.1690	0.1926**	1.0000	
% oil	–0.0670	0.0077	0.0455	0.0814	0.0648	0.0136	0.06999	–0.1195	0.5524**	1.00000

*Significant at 5% level of probability; **Significant at 1% level of probability

Source : Deshmukh *et al.* (1986)

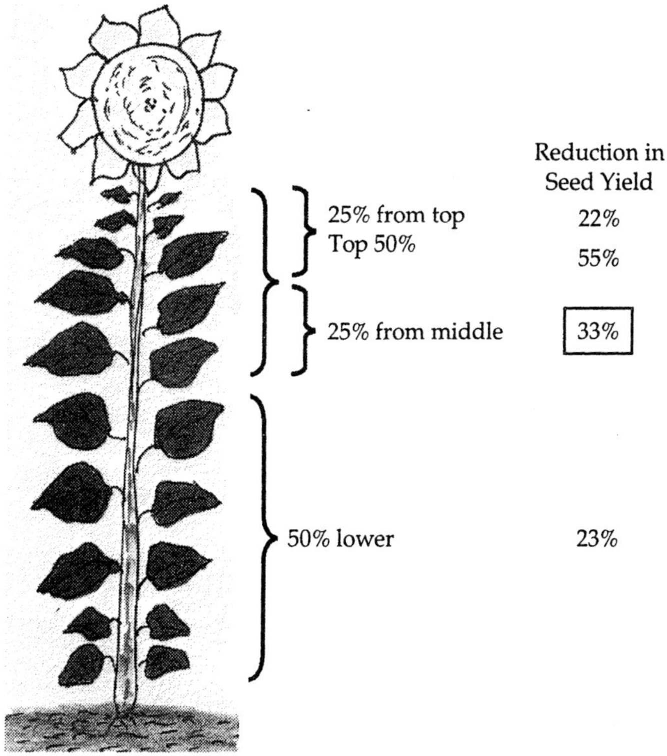

Fig. 12 : Effect of defoliation on seed yield.

Source : Srivastava *et al.*, 1977

Seed filling in the capitulum of sunflower starts from the pereferal region and moves towards the central portion in capitulum. Leaves on stem of sunflower are not contributing equally to seed filling. It has been noted that the major part of the assimilate for seed filling in sunflower is contributed by the fully developed leaves located on the top portion of the stem which have higher rate of photosynthesis. (Srivastava *et al.*,1977, Fig. 12) The older leaves at the bottom have lower rate of photosynthesis (English *et al.*, 1979) and contribute to root growth. Photosynthetic activity in the leaves which are main contributor of assimilate declines when the seeds are under filling process. Since the current photosynthesis of the top mature leaves contributes a significant amount of assimilate for seed filling, seeds remain devoid of sufficient assimilate (Srivastava and Sairam, 1983) due to reduction in photosynthetic activity of the fully mature leaves.

It has been observed that by delaying the yellowing process in leaf by foliar application of nitrogen in the form of urea and synthetic cytokinin, benzyl adenine during the post flowering period, delayed the loss in chlorophyll, RuBPcase activity and net photosynthetic carbon fixation (Table 11). There is considerable improvement in the seed filling percentage (Goswami and Srivastava, 1987). As a matter of fact this crop takes comparatively less period from seed to seed under Indian conditions compared to temperate countries like Russia, Poland, Yugoslavia, Canada etc. Higher temperature causes early maturity of this crop shortening of the total crop duration. Prolongation of active photosynthesis is likely to enhance, seed yield further and the possibility of extending at through genetic manipulation needs to be explored.

Table 11 : Seed Weight (g) per Plant at Harvest as Influenced by BA and Urea Treatments at Different Stages of Growth and Development; Average of 5 Replication in each Case

Treatments	*Time of Application (Days After Sowing)*					
	30	*60*	*90*	*60+90*	*30+60+90*	*Mean*
BA (50 ppm)	11.87	12.25	11.57	13.23	13.97	12.56
BA (100 ppm)	11.65	13.13	10.80	14.34	14.89	12.96
Urea (0.5%)	11.23	12.38	10.67	12.90	13.13	12.06
Control	11.52	11.52	11.52	11.52	11.52	11.52
CD at 5% for :						
Control × other treatments	1.01					
BA concentrations	0.554					
BA × urea	0.480					
Time of application	1.23					
BA concentration × time of application	0.91					
Treatment × time of application	0.81					

Source : Goswami and Srivastava (1987).

ROLE OF BRACT

Fig. 13 : Showing large number and size of bracts.

Ample evidences indicate that, not only leaves but other plant parts as well significantly contribute assimilates to developing seeds. The estimated photosynthetic contribution of the head and fruiting structures in different crops have been reported to be 8-23 per cent in rice (Imaizume *et al.*, 1990), 20 per cent in chickpea (Sheoran *et al.*, 1987) and 61-100 per cent in rapeseed (Sheoran *et al.*, 1991; Singal *et al.*, 1987). In cotton, bracts play a significant role in contributing photosynthates for fibre elongation (Morris, 1965) and are approximately 20 per cent as efficient as leaves in $^{14}CO_2$ assimilation (Elmore, 1973). Though, they have reduced photosynthetic capacity, they are important source of assimilates in low irradiance and water deficit environments (Wullschleger and Oosterhuis, 1990). Prasad and Srivastava (1991) have reported photosynthetic CO_2 fixation rate and light reaction in terms of photosystem I and II in bracts of sunflower alongwith leaves of different ages (Table 12 & 13).

Table 12 : Net Photosynthetic Rate (Pn) and Chlorophyll (Chl) Content in Leaves and Bracts of Sunflower. Mean of Three Replications

Leaf/bract	*Pn [μg(CO_2) m^{-2} s^{-1}]*	*Chl a*	*Chl b [g kg^{-1} (f.w.)]*	*Chl (a+b) [g kg^{-1} (f.w.)]*	*Chl a:b*
Developing leaf	273.1 ± 0.3	0.79 ± 0.16	0.51 ± 0.14	1.37 ± 0.14	1.50 ± 0.13
Fully mature leaf	520.0 ± 3.6	1.27 ± 0.15	0.80 ± 0.11	2.19 ± 0.13	1.50 ± 0.17
Senescing leaf	159.4 ± 3.1	0.86 ± 0.19	0.54 ± 0.16	1.47 ± 0.16	1.60 ± 0.14
Bract	37.8 ± 4.7	0.54 ± 0.10	0.32 ± 0.12	0.92 ± 0.15	1.70 ± 0.12

Source : Prasad and Srivastava (1991)

Table 13 : Partial Light Reactions (PS = Photosystem) in the Thylakoids Isolated from Leaves and Bracts of Sunflower

Leaf/Bract	*PS 1*	*PS 2 [meq e^- kg^{-1} (chl) s^{-1}]*	*Whole Chain [meq e^- kg^{-1} (chl) s^{-1}]*	*PS 1 : PS 2*
Developing leaf	42	156	117	0.26
Fully mature leaf	50	195	79	0.28
Senescing leaf	87	130	117	0.67
Bracts	11	91	79	0.12

Source : Prasad and Srivastava (1991).

The carbon fixation rate in bract although is less compared to fully mature leaves, however the total contribution of all the bracts becomes significant particularly at time when leaves start senescening. The bracts remain green for alonger period and contribute n seed filling. Its contribution under stress environment cannot be overlooked as bracts are less affected by water or temperature stress. Sunflower has large number of bracts (Fig. 13) and bring very close to capitulum contribute assimilate significantly to seed filling. In an experiment with defoliation of bracts have revealed a substantial reduction in seed yield (Table 14). However, the contribution of leaves is substantial because of their relatively higher surface area and net photosynthetic rates.

Sunflower has large number of bracts and being very close to the capitulum contribute assimilate significantly to seed filling. Removal of bract from head in any amount reduces head size, seed weight and 100 seed weight significant.

Table 14 : Effect of Removal of Bracts on Head Size, Seed Weight and 100 Seed Weight in Sunflower (var. EC 68413)

No.	*Treatments (Bracts Removed)*	*Time of Treatment*	*Head Diameter (cm)*	*Seed Weight % Reduction*		*100 Seed Weight % Reduction*	
T1	10 bracts	At flower opening	10.8	31.8	8.0	6.634	2.3
		10 days later	10.7	34.0	1.9	6.598	2.8
T2	20 bracts	At flower opening	10.9	28.6	17.3	6.057	10.8
		10 days later	10.8	30.9	10.7	6.388	5.9
T3	30 bracts	At flower opening	9.4	21.2	39.0	5.240	22.9
		10 days later	10.5	23.8	31.2	6.251	8.0
T4	All bracts	At flower opening	9.3	20.6	40.4	5.292	22.1
		10 days later	9.6	22.2	32.4	6.191	18.8
T5	Control No bract (removed)	–	12.6	34.6	–	6.788	–
	CD for treatments						
			at 5%	1.23	1.37	0.456	
			At 1%	1.67	1.84	NS	
			Stage	NS	NS	NS	

Source : Srivastava *et al.* (1977).

Table 15 : Percent Contribution of Bract in Seed Yield in Sunflower var. (EC 68413)

Treatments (Removal of Bracts) (%)	*Reduction in Yield (%)*
25	8
50	17
75	39
100	40

Source : Srivastava *et al.* (1977)

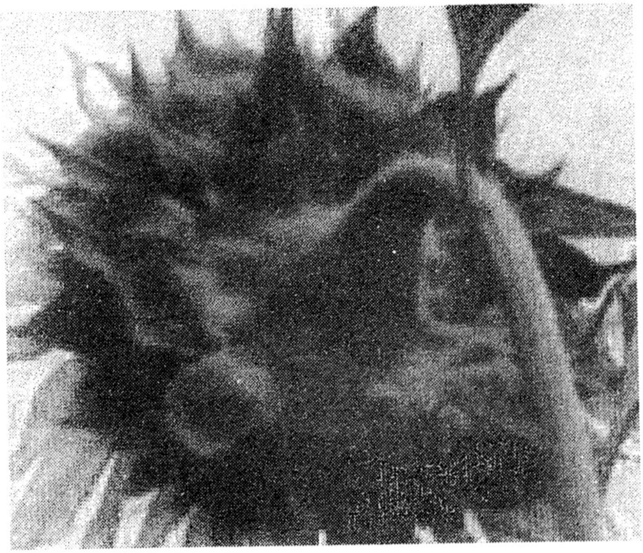

Fig. 14 : Role of Bracts in seed yield.

Fig. 15 : Branching in sunflower.

Branching is an undersirable trait in commercial sunflower production (Fig. 15). Branching of various degrees can occur in sunflower, ranging from a single stem with a large single inflorescence in cultivated types to multiple branching from axils of most leaves on the main stem in the wild species. Branch length varies from a few centimeters to a distance longer than the main stem. Branching may be concentrated at the base or top of the stem or spread throughout the entire plant. Generally, heads on branches are smaller than heads on the main stem.

Seeds from the branched flowers are normally small and seed filling takes more time causing delay in maturity. This asynchronous harvest time reduces the yield and oil content in seed.

PLANT IDEOTYPE FOR SUNFLOWER

Plant ideotype depends upon the environmental and edaphic factors. Kioezowski (1978) suggested 100 cm plant height for polish condition as ideal since under field condition, this will eliminate the risk of lodging and make harvesting easier and quicker. Jensema (1972) while comparing the yield and yield characters of maize plant and sunflower plant under same population density attributed higher yield of maize to its ear number per plant and seed weight.

Table 16 : Comparison of Yield Attributing Characters and Yield in Sunflower and Maize

Characters	*Sunflower*	*Maize*
Yield (t/ha)	2.5	5
Ear per plant	1	2.3
Yield (g)/plant	50	100
Seed (g)/plant	625	400
Weight (g)/seed	0.08	0.25
Population	50000	50000
Harvest index (%)	26	50

Source : Jensma, 1972.

It has been argued that in sunflower two head per plant may yield better but it should have synchrony in maturity. Rowson (1988) is of the opinion that attention should be paid to possibly shortening the pre-initiation phase without loss of leaf number and area by selecting for short phyllochron interval and concurrently for rapid leaf expansion rate. High leaf area, particularly during the pre anthesis period when the florets are growing to a size of competence to set seed is essential for high yield. For dry land agriculture, attention should be paid to further improving water use efficiency to result in more water being available for the maintenance of leaf area during seed filling.

Connor and Hall (1997) have shown that in sunflower crop biomass production and yield are positively coupled with solar radiation intercepted by the canopy. If an increase in yield is expected after culturing sunflower in narrow row of high densities then fixing a priory canopy parameters that could be associated with solar radiation interception is necessary to redesign the plants ideotype. Hernandez (2009) has demonstrated that with increasing light availability, steeper inclination angles in the upper canopy in a broad leaf crop plants as the sunflower become more convenient from the point of view of the photosynthetic balance. Such canopy structure allows more uniform distribution of light within the canopy and thus enables greater exposure of the photosynthesis foliar area to light.

In India, studies have been made to find the yield contributing factors under normal and abiotic stress environment. Under normal irrigated field conditions, morphological attributes such as plant height stem and head diameter, head and 1000 seed weight bear a positive and significant correlation with seed yield (Deshmukh *et al.* 1986). Similar observation has been made by Kaya *et al.* (2009) who have emphasized that plant height and other yield components such as 1000 seed weight and head diameter played an important role in determining seed yield in sunflower.

Radovan Marinkovi (2004) has reported a positive and highly significant correlation between seed yield and 1000 seed weight, whereas head diameter, volume, weight and kernel content showed a direct negative effect on seed yield per plant. Physiological characters associated with higher productivity have been attributed to higher leaf area duration (Srivastava and Goswami, 1988 and Goswami and Srivastava, 1988). English *et al.* (1979) is of the view that photosynthetic rate in all leaves decline shortly after full leaf expansion. During the seed filling process there is short availability of the photosynthate due to decline in the photosynthetic rate of most of the contributing leaves. In tropical region due to higher temperature leaf senesces process is initiated early which reduces the rate and total photosynthetic period. This ultimately affects the seed filling process. Modification of the plant development would be required to increase the productivity of sunflower. Rawson (1988) has emphasized the need to possibly shortening the pre initiation phase without loss of leaf number and area by selecting for short phyllochron interval and concurrently for rapid leaf expansion rate. High leaf area particularly during pre anthesis period when florets are growing to a size of competence to set seed is essential for high yield. Thus suggestive modification of the plant architecture and development would be necessary to develop an ideal plant to increase productivity in sunflower.

REFERENCES

Abelardo, J., Vega de la and Hall, A.J. (2002). Effect of planting dates, genotype and their interaction in sunflower yield. *Crop Sci.*, **42** : 1191-1201.

Afzalparkar, A.B. and Lakrminarayana, G. (1980). Variation in oil content and fatty acid composition with sunflower head size and shape. *J. American Oil Chem. Soc.*, **57**(3) : 105-106.

Allard, H.A. and Garner, W.W. (1940). Further observation on the response of various species of plants to length of day US Deptt. Agric. Tech. Bull. 727.

Anderson, W.K., Swith, R.C.G. and McWillium, J.R. (1978). A system approach to the adoption of sunflower to new environment, I. Phenology and development. *Field Crops Res.*, **1** : 141-152.

Anonymous (1991). Plant physiology research at IARI, Division of Plant Physiology, IARI, New Delhi, pp. 177.

Anonymous (2008). Sunflower production guidelines for the northern grains region. NSW Department of Primary Industries, Australian Oilseed Federation pp. 1-8.

Ashraf, M. and Tufail, M. (1995). Variation in salinity tolerance in sunflower (*Helianthus annuus* L.). *J. Agron. Crop Sci.*, **174** : 351-362.

Avetisyan, G.A. (1965). Bee pollination of agricultural crops. His Pehelovodstvo. P. 209-248. Moskva Kolos (In Russian).

Bakht, J., Shafi, M., Yousaf, M. and Shali, H.U. (2010). Physiology, phenology and yield of sunflower (autumn) as affected by NPK fertilizer and hybrid. *Pak. J. Bot.*, **42**(42) : 1909-1922.

Baldini, M., Vannozzi, G. Macclia, M. and Turli, M. (1996). Relationship between water and nitrogen supplied on yield and nitrogen utilization and partitioning in sunflower 412-418. Proceedings of the 14th International sunflower conference. Beijing-Shenyang, China.

Bana, R. and Srivastava, G.C. (1987). Phososynthesis leaves of sunflower (*Helianthus annuus* L.). as influenced by then position. *Sci. and Cult.*, **53** : 181-183.

Barbier, E. and Abid, M. (1966). Pollination and seed formation in sunflower. *Al. Awamia*, **21** : 63-83.

Barrett, C.F. (1954). Sunflower pollination. *Ent. Soc. Mamtoba Proc.*, **10** : 25-28.

Beaudoni, F., Lacey, D. and Napier, A. (1999). The biogenesis of the plant seed oil body : oleosin protein is synthesized by ER-bound ribosomes. *Plant Physiology and Biochemistry*, **37** : 481-490.

Bolyakina, Y. and Raikhman, L. (1999). Structural aspects of protein accumulation in developing sunflower seeds. *Russian Journal of Plant Physiology*, **46** : 340-348.

Cardon, P.V. (1922). Sunflower studies. *Amer. Soc. Agron. Jour.*, **14** : 69-72.

Chandra, S. and Anand, I.J. (1977). Association of morphological characters with seed yield and grain filling in sunflower. *Crop Improvement*, **4**(2) : 170-176.

Chimenti, C.A., Pearson, J. and Hall, A.J. (2002). Osmotic adjustment and yield maintenance under drought in sunflower. *Field Crop Research*,**75**(2-3) : 235-246.

Connor, D. and Sadras, V. (1992). Physiology of yield expression in sunflower. *Field Crops Res.*, **30** : 333-389.

Connor, D.J. and Hall, A.J. (1997). Sunflower physiology. In: Schneiter A.A. (ed.) sunflower technology and production. American Society of Agronomy Madison W1 USA pp. 113-182.

Connor, D.J. Jones, T.R. and Palta, J.A. (1985). Response of sunflower to strategies of irrigation I growth, yield and efficiency of water use. *Field Crop Res.*, **10** : 15-36.

Connor, D.J., Palta, J.A. and Jones, T.R. (1985). Response of sunflower to strategies of irrigation III. Crop Photosynthesis and respiration. *Field Crop Res.*, **12** : 281-293.

Deshmukh, P.S. and Srivastava, G.C. (1983). Nitrate reductase activity in selection to yield in sunflower. *Indian J. Pl. Physiol.*, **26**(2) : 143-1470.

Deshmukh, P.S., Srivastava, G.C. and Pando, S.B. (1988). Influence of moisture stress and cycocel (2 chloroethyl trimethyl ammonium chloride) on prolien accumulation and intrate reductase activity in sunflower. *New Botanist*, **15**(2-3) : 117-121.

Deshmukh, P.S., Srivastva, G.C. and Tomar, D.P.S. (1986). Effect of environmental factors on correlation coefficient between morphological parameters of yield in sunflower (*Helianthus annuus* L.). *Indian J. Pl. Physiol.*, **29** : 345-350.

Elmore, C.D. (1973). Contributions of the capsule wall and bracts to the developing cotton fruit. *Crop Sci.*, **13** : 751-752.

English, S.D., McWilliam, J.R., Smith, R.C.G. and Davidson, J.L. (1979). Photosynthesis and partitioning of dry matter in sunflower. *Aust. J. Pl. Physiol.*, **6** : 149-164.

Escalante, J.A., Rodriguez, M.T., Dettaro, A. and Fiereres, E.C. (1998). Acquisition, partitioning and remobilization of nitrogen and then relationship to seed yield inmediterraneam sunflower. *Hetia*, **21** : 81-94.

Espelund, M., De Bedout, J.A., Outlaw, W.H. and Jakobsen, K.S. (1995). Environmental and hormonal regulation of barley late-embryogenesis-abundant (*Lea*) mRNAs is via different signal transduction pathways. *Plant Cell Environ.*, **18** : 943-949.

Evans, J.R. (1989). Photosynthesis and nitrogen relationship in leaves of C_3 plants. *Oecologia*, **78** : 9-19.

Farquhar, G.D., Dubbe, D.R. and Raschke, K. (1978). Gain of the feedback crop involving carbon dioxide and stomata. *Plant Physiol.*, **62** : 406-412.

Firscher, R.A. and Kertenz, Z. (1976). Harvest index in spaced population and grain weight in micropltos as indicators of yielding ability in spring wheat. *Crop Sci.*, **16** : 55-59.

Francis, Flenet, Kiniry, J.R., Board, J.E., Westgate, M.E. and Reicosky, D.C. (1996). *Agron. J.*, **88**(2) : 185-190.

Francois, L.E. (1996). Salinity effect on form sunflower hybrids. *Agronomy J.*, **88** : 215-219.

Free, J.B. (1964). The behaviour of honey bees on sunflower (*Helianthus annuus* L.). *J. Appl. Eco.*, **1**(1) : 19-27.

Free, J.B. and Simpson, J. (1964). The pollination requirement of sunflower (*Helianthus annuus* L.). *Empire J. Expt. Agr.*, **32**(128) : 340-342.

Gasev, N.A. and Gordon, L.H. (1968). Fiziologiga vodoobmena 1 ustojeivosti rastenij. Kazan (in Russian).

Goswami, B.K. (1985). Studies on photosynthesis and nitrogen metabolism in relation to growth and yield of sunflower (*Helianthus annuus* L.) as influence by benzyladenine and urea. Ph.D. Thesis IARI, New Delhi. p. 155.

Goswami, B.K. and Srivastava, G.C. (1987). Modification of leaf senescence and seed in sunflower (*Helianthus annuus* L.) by benzyladenium and urea. *Indian J. Pl. Physiol.*, **30** : 337-343.

Goswami, B.K. and Srivastava, G.C. (1988). Effect of foliar application of urea on leaf senescence and photosynthesis in sunflower (*Helianthus annuus* L.). *Photosynthetica*, **22**(1) : 99-104.

Goyne, P.J., Woodruft, D.R. and Churchett, D. (1977). Prediction of flowering in sunflowers. *Aust. J. Exp. Agric. Anim. Husb.*, **17** : 475-481.

Grassini, P., Indaco, G.V., Lopez Perrira, M., Hall, A.J. and Trapani, N. (2007). Responses to short term water logging during grain filling in sunflower. *Field Crop Res.*, **101**(3) : 352-363.

Haverkkort, A.J., Waart, M.V.D. and Bodlaender, K.B.A. (1990). The effect of early drought stress on number of tubers and stolons of potato in controlled and field conditions. *Potato Res.*, **33** : 89-96.

Hernandez, L.F. (2009). Leaf inculcation and light interception in the sunflower (*Helianthus annuus* L.). Importance of the petioles mechanical and structural properties 6th Plant Biomechanics Conference Cayenne, Nov. 16-21.

Hiller, E.A. and Clark, R.N. (1971). Stress day index to characterize effects of water stress on crop yields. Transactions of the ASAE, pp. 757-761.

Hocking, P.J., Randall, P.J. and Pinkerton, A. (1987). Sulfur nutritional of sunflower (*Helianthus annuus* L.) as affected by nitrogen supply. Effect on vegetative growth, the development of yield components and seed yield and quality. *Field Crop Res.*, **16**(2) : 157-175.

Houret Jessica (2002). Practical Herb Garden. Pub. Hermes House. London.

Huang, A.H. (1992). Oil bodies and oleosins in seeds. Annual Review of Plant Physiology and Plant Molecular Biology. **43** : 177-200.

Imaizumi, N., Usuda, H., Nakamoto, H. And Ishihara, K. (1990). Changes in the rate of photosynthesis during grain filling and the enzymatic activities associated with the photosyntheic carbon meabolism in rice panicles. *Plant Cell Physiol.,* **31** : 835-843.

Ingram, J. and Bartels, D. (1996). The molecular basis of dehydration tolerance in plants. *Annu. Rev. Pl. Physiol. Pl. Mol.Biol.,* **47** : 377-403.

Jensma, J.R. (1972). Sunflower bloom as a new oilseed crop. *Agric. Digest,* **25** : 51-54.

Jeong, M.J., Park, S.C., Kwon, H.B. and Byun, M.O. (2000). Isolation and characterization of the gene encoding glyceraldehydes-3-phosphate dehydrogenase. *Biochem. Biophys. Res. Commun.,* **278** : 192-196.

Kalton, R.R. (1951). Efficiency of various bagging materials for efficiency self fertilization of sunflower. *Agron. J.,* **43**: 328-331.

Katarji, N., Van Horn, J.W., Hamdy, A. and Mastrorilli, M. (2003). Salinity effect on crop development and yield, analysis of salt tolerance according to several classification methods. *Agric. Water Manag.,* **62** : 37-66.

Kaya, Y., Evci, G., Durak, S., Pekcan, V. and Gucer, T. (2009). Yield components affecting seed yield and their relationships in sunflower (*Helianthus annuus* L.). *Pak. J. Bot.,* **41**(5) : 2261-2269.

Khanna, K.R. (1972). Factors affecting the production of filled seeds in sunflower. *Euphytica,* **21**(2) : 384-387.

Kiani, S.P., Grieu, P., Maury, P., Hewezi, T., Gentzbittel, L. and Sarrafi, A. (2007b). Genetic variability for physiological traits under drought conditions and differential expression of water stress-associated genes in sunflower (*Helianthus annuus* L.). *Theoretical and Applied Genetics,* **114**(2) : 193-207.

Kim, J.Y., Mahe, A., Brangeon, J. and Prioul, J.L. (2000). A maize vacuolar invertase, IV2, is induced by water stress. Organ/tissue specificity and diurnal modulation of expression. *Plant Physiol.,* **24** : 71-84.

Kloezowski, Z. (1978). Morphological and physiological ideotype of sunflower in Poland. Proceedings of meeting on morphological and physiological ideotype of oil crops held at Uppsala, Sweden.

Kogan, M.J., Dalcol, I., Gorostiza, P., Lopez-Lglsias, C., Pons, R., Pons, M., Sanz, F. And Giratt, E.(2000). Super molecular properties of the protein rich xzem-N terminal domin. *Biophysical J.,* **83** : 1194.

Kurvits, A. and Kirby, E.A. (2007). The uptake of nutrients by sunflower plants (*Helianthus annuus* L.) growing is a continuous flowering cultivar system, supplied with nitrate or ammonium as nitrogen source. *J. Plant Nutrition and Soil Sci.,* **143**(2) : 140-149.

Levy, D. (1992). The response of potatoes (*Solanum tuberosum* L.) to salinity : plant growth and tuber yields in the arid desert of Isreal. *Ann. Appl. Biol.,* **120** : 547-555.

Lindstrom, L.I., Friedrich, P., Hernandez, L.F. and Pellegrini, C.N. (2002). Embryo cell dynamics during early development of the sunflower fruit. In: First Latinamerican Plant Physiology Congress Procs. Punta del Este, Uruguay, p. 276.

Lopez Pereira, M., Sadras, V. and Trapani, N. (1999). Genetic improvement of sunflower in Argentina between 1930 and 1050. I. Yield and its components. *Field Crops Res.,* **62** : 157-166.

Lopez-Perura, M., Sadras, V. and Trapani, N. (2000). Genetic improvement in sunflower in Argentina between 1930 and 1995. Part III. Dry matter partitioning and achene composition. *Field Crop Res.,* **67** : 215-225.

Lynch, D.R. and Tai, G.C.C. (1989). Yield and yield components response of eight potato genotypes to water stress. *Crop Sci.,* **29** : 1207-1211.

Manmohan, M. (1995). Rubisco in sunflower (*Helianthus annuus* L.) as affected by water stress. Ph.D. thesis, IARI, New Delhi. pp. 72.

Marinkovic, R., Joksimovic, J., Jovanovic, D. and Jeromela, A.M. (2004). Assessment of components genetic variance of some quantitative traits in sunflower (*Helianthus annuus* L.). *Bulgarian J. Agril. Sci.,* **10**(5) : 539-547.

McWilliam, J.R., English, S.D. and McDougall, G.N. (1974). The effect of leaf age and position on photosynthesis and the supply of assimilates during development in sunflower. In 6th International Sunflower Congress Proc. 1974.

Mitra, S. and Srivastava, G.C. (1993). Effect of altered assimilate level by dark starvation on SPS activity and photosynthesis in sunflower (*Helianthus annuus* L.). *Indian J. Pl. Physiol.,* **37**(3): 160-163.

Montemurrod, F. and Giorgio, D.D. (2005). Quality and nitrogen use efficiency of sunflower grown at different nitrogen levels under Mediterranean condition. *Journal of Plant Nutr.,* **28**(2) : 235-350.

Moons, A., Prinsen, E., Bauw, G. and Van Montagu, M. (1997). Antagonistic effects of abscisic acid and jasmonates on salt stress inducible transcripts in rice roots. *Plant Cell,* **9** : 2243-2259.

Morris, D.A. (1965). Photosynthesis by capsule wall and bracteoles of cotton plants. *Emp. Cotton Grow Rev.,* **42** : 49-51.

Muralidharudu, Y., Murthy, I.Y.L.N., Reddy, K.P.C., Reddy, B.N. and Chandranath, M.T. (2003). Response of sunflower (*Helianthus annuus* L.) to phosphorus application in vertisols. *Helia,* **26** : 147-154.

Muralidharudu, Y., Ravishankar, G., Hebbara, M. and Patil, S.C. (1999). Genotypic variation in sunflower (*Helianthus annuus* L.) for salt tolerance. *Indian J. Agric. Sci.,* **69** : 362-365.

Murphy, D.J. (2001). Biogenesis and functions of lipids bodies in animal, plant and microorganisms. Progress in Lipid Research (2001).

Nepomuceno, A.L., Stewart, J.M.C.D., Oosterhuis, D., Turley, R., Neumaier, N. and Farias, J.R.B. (2000). Isolation of a cotton NADP(H) oxidase homologue induced by drought stress. *Pesqui. Agropecu. Bras.,* **35** : 1407-1416.

Noetzel, D.M. (1968). Insect pollination results on sunflower. Dept. Ent. N. Dak State University, Fargo, pp. 108-112 (Mimeographed).

Orchard, P.W. and Jessop, R.S. (2005). The response of sorghum and sunflower to short term waterlogging. I effect of stage of denlopust and duration of water logging on growth and yield. *Plant and Soil,* **81**(1) : 119-132.

Perniola, M. (1994). Ecophysiological parameters and water relations of sweet sorghum, cotton and sunflower during drought cycle. International Conference on land water resource management in the medeteaean region. Valenzano Bari Int. 4-8 Sept., **5** : 97-115.

Ploschuk, E. and Hall A. (1997). Maintenance respiration coefficient for sunflower achenes is less than that for the entire capitulum. *Field Crops Res.,* **49** : 147-157.

Ploschuk, E.L. and Hall, A.J. (1995). Capitulum position in sunflower affects grain temperature and duration of grain filling. *Field Crop Res.,* **44**(2-3) : 111-117.

Prasad,N.K.(1990). Some aspects of photosynthesis in sunflower (*Helianthus annus,* L) under moisture stress condition. Ph.D. Thesis, IARI, Delhi p. 84.

Prasad, N.K. and Srivastava, G.C. (1990). Photosynthetic acclimation to water stress in sunflower (*Helianthus annuus* L.). *Indian J. Pl. Physiol.,* **33**(4) : 318-326.

Prasad, N.K. and Srivastava, G.C. (1991). Pantial light reaction and carbon fixation in relation to leaf ontogeny and in bracts of sunflower (*Helianthus annuus* L.). *Photosynthetica,* **25**(4) : 499-503.

Prasad, N.K., Mitra, S. and Srivastava, G.C. (1989). Photosynthetic rates in relation to leaf position in sunflower (*Helianthus annuus* L.). *Indian J. Pl. Physiol.,* **32** : 383-386.

Rawson, H.M, Dunstone, R.L., Long, M.J. and Begg, J.E. (1984). Canopy development, light interception and seed production in sunflower as influenced by temperature and radiation. *Australian Journal of Plant Physiology*, **11** : 255-265.

Rawson, H.M. (1988). Can productivity in sunflower be increased by changed plant development. Proc. Intrnt. Congr. Pl. Physiol., New Delhi, pp. 15-20.

Rawson, H.M., Constable, G.A. and Howe, G.N. (1980). Carbon production of sunflower cultivars in field and controlled environments II. Leaf growth. *Austr. J. Pl. Physiol.*, **7** : 575-586.

Rodole, J.I. (1968). The natural way to better eye sight. N.V. Publication, New Delhi.

Russell, W.A. (1953). A study of the inter relationship of seed yield, oil content and other agronomic characters with sunflower by inbreed lines and their top crosses. *Can. J. Agric. Sci.*, **33** : 291-314.

Sanduleac, E. (1960). Insect Pollination of the Sunflower. *Lucr. Stint. Stat. Cent. Seri. Apic.*, **2** : 209-218.

Schelotto, B. and Pereyras, N.L. (1971). An evaluation of the economic significance of pollinating sunflower with bees. *Cieciay Abejas*, **1**(2) : 7-25 (In Spanish, English Summary).

Sheoran, I.S., Singal, H.R. and Singh, R. (1987). Photosynthetic characteristics of chickpea (*Cicer arietinum* L.) pod wall during seed development. *Indian J. Expt. Biol.*, **25** : 843-847.

Sheorn, I.S., Sawhney, V., Babbar, S. and Singh, R. (1991). In vitro fixation of CO_2 by attached pods of *Brassica compestris* L. *Ann. Bot.*, **67** : 425-428.

Sinclair, T.R. and Licdlow, M.M.(1985). Who taught plants thermo-dyonamics. The unfulfilled potential of plant water potential. *Austr. J. Pl. Physiol.*, **12** : 213-217.

Singal, H.R., Sheoram, I.S. and Singh, R. (1987). Photosynthetic carbon fixation characteristics of fruiting structures of *Brassica camestries*. L. *Plant Physiol.*, **83** : 1043-1047.

Skoric, D. (2009). Sunflower breeding for resistance to abiotic stress. *Hetia*, **32** : 1-16.

Srivastava, G.C. and Goswami, B.K. (1988). Influence of benzyle-adenine on leaf senescence and photosynthesis in sunflower (*Helianthus annuus* L.). *J. Agron. Crop Sci.*, **161** : 23-29.

Srivastava, G.C. and Sairam, R.K. (1983). Physiological studies on seed setting in sunflower. I Photosynthesis and nitrate assimilation. *Indian J. Pl. Physiol.*, **26** : 370-377.

Srivastava, G.C., Deshmukh, P.S. and Tomar, D.P.S. (1977). Significant of leaf orientation and bract in seed yield and sunflower. *Indian J. Plant Physiol.*, **20** : 151-156.

Srivastava, G.C., Deshmukh, P.S. and Tomer, D.P.S. (1977). Significance of leaf orientation and bract in seed yield in sunflower. *Indian J. Pl. Physiol.*, **20** : 151-156.

Tanksley, S.D. (1993). Mapping polygenes. *Annual Review of Genetics*, **27** : 205-233.

Thavaprakash, N., Senthilkumar, G., Sivakumar, S.D. and Rajan, M. (2003). Photosynthetic attributes and seed yield of sunflower (*Helianthus annuus* L.) as influenced by different levels and ratio of nitrogen and phosphorus fertilizers. *Acta Aro. Hingaric*, **51**(2) : 149-155.

Tomar, D.P.S., Deshmukh, P.S. and Sinha, S.K. (1979). Importance of sepals in fruits and seed development in linseed. *Euphytica*, **28** : 739-745.

Turhan, H. and Ayaz, C. (2004). Effect of salinity on seedling emergence and growth of sunflower (*Helianthus annuus* L.) cultivars. *Indian Journal of Agriculture and Biology*, **6**(1) : 149-152.

Uno, Y., Furihata, T., Abe, H., Yoshida, R., Shinozaki, K. And Yamaguchi-Shinozaki, K. (2000). Arabidopsis basis leucine zipper transcription factors involved in an abscisic acid-dependent signal transduction pathway under drought and high salinity conditions. *Proc. Natl. Acad. Sci. USA*, **97** : 11632-11637.

Virgona, J.M. and Farquhar, G.D. (1996). Genotypic variation in relative growth rate and carbon isotope discrimination in sunflower is related to photosynthetic capacity. *Aust. J. Pl. Physiol.*, **23** : 227-236.

Virgona, J.M., Hubick, K.T., Rawson, H.M., Farquhar, G.D. and Downcs, R.W. (1990). Genotypic variation in transpiration efficiency, carbon isotope discrimination and carbon allocation during growth in sunflower. *Austr. J. Pl. Physiol.*, **17** : 207-214.

Warren-Wilson, J. (1966). High net assimilation rates of sunflower plants in an arid climate. *Ann. Bot.*, **30** : 745-751.

Wei, J., Tirajoh, A., Effendy, J. and Plant, A. (2000). Characterization of salt-induced changes in gene expression in tomato (*Lycopersicon esculentum*) roots and the role played by abscisic acid. *Plant Sci.*, **159** : 135-148.

Wong, S.C., Cowan, I.R. and Farquhar, G.D. (1979). Stomatal conductance correlates with photosynthetic capacity. *Nature*, **282** : 424-426.

Wullschleger, S.D. and Oosterhmis, D.M. (1990). Photosynthetic and respiratory activity of fruiting forms with in the cotton canopy. *Plant Physiol.*, **94** : 463-469.

Yegappan, I.M., Paton, D.M., Gates, C.T. and Muller, W.J. (1982). Water stress in sunflower (*Helianthus annuus* L.) 3 responses of cypsela size. *Ann. Bot.*, **49** : 69-75.

5

INDIAN MUSTARD

Order	:	Brassicales
Family	:	Brassicaceae
Genus	:	*Brassica*

Fig. 1 : Indian mustard plant.

Rapeseed mustard, one of the species of *Brassicas* (Fig.1) is widely cultivated in Indian subcontinent. *Brassica* crops are among the oldest cultivated plants known to humans with written records dating back to Ca B 1500 BCE (Prakash, 1980) and archaeological evidence of its importance dating back to 5000 BCE (Yan, 1990). *Brassica rapa* is believed to be the oldest species and seems to have the widest distribution historically. At least 2000 years ago, it was distributed from northern Europe to China and Korea, with a primary center of diversity in the Himalayan region (Hegde, 1976). *Brassica napus* is believed to have developed in the Mediterranean area where the wild forms of its ancestral species were sympatric. Production of oilseed *B. napus* probably started in Europe during the middle ages where its oil was used as a lamp oil.

In addition to *B. napus* L. and *B. rapa* L., *Brassica* includes cultivated species such as *B. carinata* Braun, *B. nigra* Koch, and *B. oleracea* L. The four most widely cultivated species *B. juncea, B. napus, B. oleracea* and *B. rapa* are all highly polymorphic and include oilseed crops, root crops and vegetables crops such as Chinese cabbage, Broccoli, and Brussels sprouts. The relationships among the cultivated species as presented in Fig. 2 were first clarified by Morinaga (1934) and verified by "U" (1935). *B. napus* (2n=38, AACC), *B. juncea* (2n=36, AABB), and *B. carinata* (2n=34, BBCC) are amphidiploids species resulting from combining chromosome sets of the low chromosome number species *B. nigra* (2n=16, BB), *B. oleracea* (2n=18, CC), and *B. rapa* (2n = 20, AA).

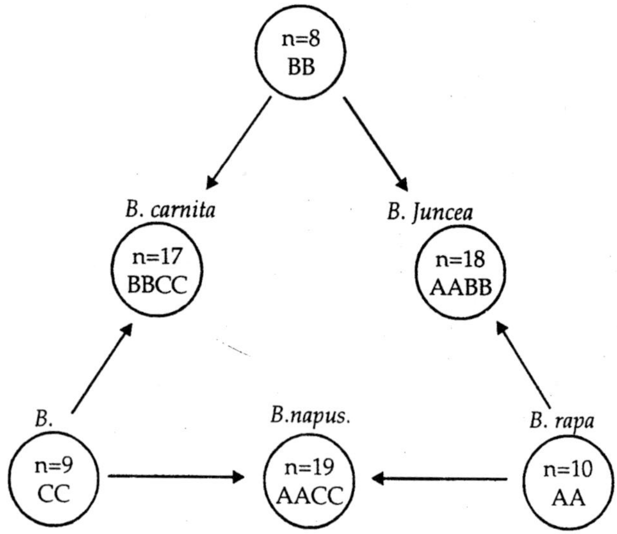

Fig. 2 : The "Triangle of U" representing the genomic relationships among *Brassica* species.

Basically, *Brassica* is a low temperature crop and grown in tropical and subtropical countries. Breeding for high temperature tolerance of these crops can have an advantage of its early sowing for increased harnessing soil moisture stored in the root zone due to mansoon rains. India although occupies first position in the acreage and production of mustard in the world, the average yield per hectare is low as compared to other countries where mustered is grown.

In India, rapeseed mustered is grown as a dry season crop (winter) in sub tropical area mainly on stored soil moisture from the previous wet season. Crop thus flowers and ripen under increasing water stress condition. Oilseed *Brassicas* are listed in four with an indication of the ecological niche which each one occupies (Bunting, 1986).

BRASSICA RAPA

Fig. 3 : *Rapa* plant in flowering stage.

Brassica rapa L. (syn. *Brassica campestries* L.) (Fig. 3) seems to have grown naturally from the West Mediterranean region to Central Asia and is still present throughout this area and in general associated with weedy habitats. The earliest reference pertains to yellow sarson in ancient Sanskrit literary works such as the Upnishadas where it is referred as Siddhartha (Prakash, 1961). *Brassica rapa* has the widest distribution with secondary entries of diversity in Europe, western Russia, Central Asia and the near East (Vavilov, 1949). Its wide availability made it probably the first domesticated *Brassica* crop. The wide range of sub species in *B. rapa* is bellowed to have resulted from varied selection pressure in different geographic regions.

BRASSICA NAPUS

The highest yield potential shown by this species, at least in favourable environment, is probably due to its allopolyploid origin. Winter types are generally grown in Europe, and spring types in other area. The amphiploid species *Brassica napus* (oilseed rape, Canola) (Fig. 4) is a globally important oil crop yielding food, biofuels and inducstrial compounds such as lubricants and surfactants. Identification of the likely ancestors of each of the two genomes (designated A and C) found in *B. napus* would facilitate incorporation of novel alleles from the wider *Brassica* genepool in oilseed rape crop genetic improvement programmes.

Fig. 4 : *B. napus* plant in flowering stage.

BRASSICA JUNCEA

Brassica juncea (Indian mustard) (Fig. 5) is the major oilseed in Asia. Interest has been generated in Canada (Love *et al.*, 1991) and Australia (Oram and Kirk, 1992) recently since low erucic acid and low glucosinolate types are becoming available. It has been shown to be more heat and drought tolerant than the other species, with a range of contributing characters. For example, the better germination on drying soils appears to be due to the higher concentration of mucilage in the testa of *B. juncea* seeds than in *B. napus*, allowing water attraction and retention (Oram and Kirk, 1992).

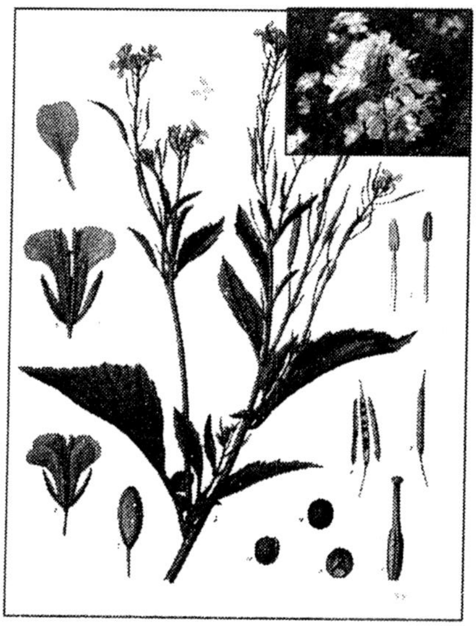

Fig. 5. *Juncea* in flowering stage.

BRASSICA CARINATA

Brassica carinata (Ethiopian mustard, Abyssinian mustard) (Fig. 6) is a member of the Triangle of U species (U, 1935) in the agriculturally significant *Brassica genus*. It has 34 chromosomes with genome composition BBCC, and is thought to result from an ancestral hybridization event between *Brassica nigra* (genome composition BB) and *Brassica oleracea* (genome composition CC)

(Prakash and Hinata, 1980). Although *B. carinata* is cultivated as an oilseed crop in Ethiopia (Alemayehu and Becker, 2004) it has generally high levels of undesirable glucosinolates and erucic acid (Getinet *et al.*, 1997), marking it a poor choice for general cultivation as an oilseed crop in comparison to the closely related *Brassica napus* (canola). It is drought resistant (Kumar *et al.*, 1984), plus resistance to pod shattering and disease (Alonso *et al.*, 1991). In India, Malik (1990) showed that this species performed better under saline conditions and late sowing. Its tropical origin may allow it to tolerate high temperatures during the reproductive period better than the other species.

Fig. 6 : *Carinata* in flowering stage.

There is enormous variation in all the species in terms of yield characters (Table 1&2) and the scope for inter specific crossing. Other genetic transfers provides the potential for breeding and selection to make the best use of physiological and other characteristics wherever they are found in the *Brassica* gene pool.

Table 1 : Yield Contributing Characters of *Brassica* Species on per Plant Basis at Plant Population of 11 Plants/m^2

Species	*Pod Number*	*1000-Seed wt. (g)*	*Seed yield g/Plant*	*Harvest Index*
Brassica juncea	585	6.26	36.6	31
B. campestris	431	3.74	19.8	28
B. napus	359	3.62	21.9	23
B. carinata	570	3.11	24.81	20

Source : Bhargava and Tomar (1988)

Table 2 : Reproductive and Seed Attributes of Four *Brassica* Species

Species	*Day to 50% Flowering*	*Duration of Pod Maturation in days*	*Ovule No. at Anthesis*	*Seed No. Per Pod*	*Seed Wt. Per Pod (mg)*
Brassica juncea	55	70	11	11	60
B. campestris	60	80	23	23	90
B. napus	85	60	23	22	90
B. carinata	115	30	16	16	50

Source : Anonymous (1991)

The flowering in mustard changes under different photoperiodic treatments (Table 3). Days to appearance of first

Table 3 : Effect of Photoperiods on Days to Appearance of First Flower in Four *Brassica* Ecotypes

Ecotype	*Short-day (8 hr)*	*Natural-day (10-12 hr)*	*Long-day (24 hr)*
B. campestris			
Toria	24	24	29
Brown sarson	65	46	49
Yellow sarson	65	59	59
B. juncea			
Mustard	–	53	56

Source : Bhargava and Tomar (1988)

flower decreases as the length of the photoperiod increases in brown and yellow sarson suggesting a quantitative long day behaviour of plants. In Toria however, flowering occurred early and simultaneously in short day and normal day but with slight delay under long days indicating its photoinsensitivity (Bhargava and Tomar, 1988).

In order to facilitate inter specific crossing in *brassica* species Bhargava and Tomar (1991) made an attempt to initiate flowering by providing different combinations of photoperiod and temperature and recorded that under controlled environment with 16 hours photoperiod and 27/17°C day/night temperature, all *Brassica* species (*B. rapa, B. carinata, B. napus, B. juncea* and *B. campestris*) can be made to flower synchroneously in 23-28 days after sowing. In field condition, *Brassica* species takes 23 to 110 days for flowering after sowing and thus making crossing problem. There findings have provided ways to make all the five species flower within a reasonable period.

GERMINATION AND GROWTH OF THE PLANT

Productivity significantly depends on germination and vigour of the plant to a large extent. Temperature and aeration of the soil affects the rate of seed germination. Seed that is fully mature on the mother plant is observed to have faster radicle emergence than immature seed. Larger seed, for example from main stem or upper branches, produces larger cotyledons and hence show more vigorous seedlings (Leterme, 1988b). Major (1977) graded out three sizes of seeds that showed greater seedling vigour from the large seeds, but with no effect on final population or yield. Studies by Chitra Devi *et al.* (1988) have shown that seed size and storage period had significant effect on the seed quality, this advantage however may diminish in the field where inter-plant competition exists and plants from a large and small seed may eventually attain similar yield. Mendham *et al.* (1981b) also showed an early benefit to growth from the larger seeds, which only affected later crop performance when plant size at flowering was limited by late sowing.

Fig. 7 : Root of Brassica plant.

On the other hand Elliot (2007) has found that shoot weight after 14 and 21 days, increased as the seed weight increased. The report of canola council of Canada also indicated that the size of seed can affect dry matter accumulation in seedlings. Large seed size had almost twice as much dry matter accumulation after 35 days as small seed size. The council classified large seed as 3-3.6 g/1000 seeds, medium size seed as 2.7-3 g/1000 seed and small size seed as 1.5-2.29/1000 seeds. Seeds have been graded in to three categories on the size based on sieving through grading screen. Small (1.4 mm to 1.6 mm), medium (1.6 to 1.8 mm) and large seeds (1.8 mm to 2.00 mm). Seed size thus has effect on growth of the seedling.

Root growth is extremely important particularly under dry conditions. Although studies of root growth is very important but it is a very cumbersome processes. Rapid root growth after establishment consists of tap root extension vertically, growth of the secondary roots laterally, and then laying down of reserves, principally in the tap root (Fig. 7). Most studies of roots include only those reasonably accessible to digging, to about 30 cm or where root activity is inferred from water extraction patterns. Few workers have taken the trouble of measuring roots, either from extracted cores or *in situ*. Almond *et al.* (1984) dug pits to 1.5 m and measured root distribution pattern on a grid to show the effect of cultivation method. Root numbers were distributed 27 to 52% in the top 10 cm, 14 to 26% from 10 to 20 cm, down to 1% or less below 1 m, again being close to their maximum at the end of flowering.

The growing point or apex of the plant produces leaf initials in a helical arrangement. Leaf initiation rates (plastochron) are faster than appearance rates (phyllochron), so leaf primordial accumulate around the apex. Each leaf primordium has the potential to develop a further meristem in its axil.

The leaf growth and the number are under the phenological development however there is considerable variation amongst species. In one planting experiment date with four species of *Brassica, B. camperstris* (cv. Pusa Kalyani) *B. juncea* (cv. Varuna), *B. napus* (cv. BO 706) and *B. carinata* (cv. Tall-1) sown at five dates, where photoperiod declined from 11.3 to 10.2 hr and temperature from 24 to 15°C. Nanda *et al.* (1995) reported that planting dates had a big effect on the duration to the appearance of the first leaf with the range being from 10.8 (sowing 1) to 22 days after sowing (sowing 5), equivalent to a delay of 1.35 days for each 1°C reduction in mean temperature. Base temperatures for this phase were approximately 5°C for all species. Planting date had no significant effect on the rate of appearance of subsequent leaves which ranged between 2.68 (date 1) and 2.84 days per leaf (date 3) and averaged equivalent to 0.37±0.003 leaves per day. However, rates of leaf appearance differed over a 35% range amongst species, with new leaves appearing every 2.33±0.07 days (*B. campestris*, 2.71±0.12 (*B. juncea*), 2.90±0.09 (*B. carinata)*, and 3.18±0.08 days (*B. napus*). Upper leaves emerged between 33 and 48% faster than lower leaves depending on species, and this was most apparent at the final two sowings, and particularly in *B. napus*.

Final leaf number declined with later sowing at a rate ranging between 0.186 (*B. campestris*) and 0.112 leaves per day (*B. juncea*). Absolute numbers differed amongst species from 35

(*B. carinata,* sowing 1) to 12 (*B. campestris,* sowing 5), but were positively correlated with days from sowing to floral initiation.

Fig. 8 : Sequential arrangement of flowers on stem.

Area per leaf increased with leaf position reaching a peak in the leaf that was emerging at approximately the time of floral initiation with *B. napus* leaves on average increasing fastest (100 cm^2 per leaf position) and *B. carinata* slowest (60 cm^2 per leaf). Planting date affected areas of individual leaves largely through changing the timing of floral initiation and by consequence the degree of advancement of the leaf area profiles (Nanda *et al.,* 1995).

Once requirements for vernalization or photoperiod have been met, the first sign of flower initiation will be a swelling in the axil of one of the leaf primordial at the apex. Successive primordia then develops from the axillary rather than leaf meristems, becoming the flowers of the main stem or raceme (Fig. 8). Once the main stem flower buds are forming, axillary buds lower down will then being to develop sequentially into the primary branches in a basipetal direction. At the main stem apex, floral development continues acropetally with the sepals on each bud becoming obvious, and then peduncles. Smith and Scarisbrick (1990) describe development of sepals, stamens, petals and gynoecium, the latter three occurring at about the time that buds become visible without dissection. Ovules begin to differentiate in the ovary, and flower development is completed during the yellow bud stage.

The two key points in development are inflorescence initiation and flowering. The number of leaf initials produced before initiation has a major influence on leaf area and hence photosynthetic potential of the crop, which is realized when all the main stem leaves expand by flowering time. At flowering the yield potential is set, as a balance between vegetative growth and the potential number of flowers, pods and seeds. Hence the timing of the key points is crucial to the match between genotype and environment.

DEVELOPMENT OF YIELD COMPONENTS

Brassica is indeterminate crop. While comparing individual pods with their characters on terminal raceme it revealed that changes in pod weight, seed weight, 1000 seed weight and seed number per pod have a curvilinear decrease with increasing node position (Fig. 9). Marked decline in pod and seed weight towards the apex of the raceme is mainly due to the diminishing assimilate supply. The presence of unfilled sinks on more apical nodes may indicate scope for improving yield via increase in assimilate supply by reducing pod number and synchrony of flowering on different order branches (Bhargava and Tomar 1988).

Removal of 3 to 4 basal primary branches as and when they appear increased the seed yield per plant of mustard cvs Pusa Bold by 36% and Varuna by 75% over control. The results suggest that improvement in yield by suppressing branching on basal portion of the plants should be possible.

Fig. 9 : Arrangement of siliqua on stem.

The yield in mustard is decided by the total number of siliqua on the plant and seed number and weight of the seed per siliqua (Table 4). Srivastava and Balkrishna (2003) have shown a good correlation between siliqua per plant and seed yield per plant.

Table 4 : Simple Correlation Coefficient Between Yield and Yield Attributes in Mustard. Based on the Observations of Three Years and Two Genotypes (*Brassica juncea* cv. Vaibhav and Kranti)

Characters	*TDMP Per Plant*	*Speed Yield Per Plant*	*Speed Yield Per Siliqua*	*Speed Number Per Siliqua*	*1000 Seed Weight*	*Harvest Index*
Seed yield per plant	0.81					
Seed weight per siliquae	0.42	0.33				
Seed number per siliquae	0.08	-0.06	0.74			
1000 seed weight	0.1	0.57	0.54	0.15		
Harvest index	-0.26	0.61	-0.05	-0.17	0.22	
Siliquae per plant	0.46	0.73	-0.32	-0.57	0.16	0.61

Source : Srivastava and Balkrishna (2003).

PHOTOSYNTHESIS AND SOURCE SINK RELATIONSHIP

Mustard belongs to C_3 group of plants. It is characterized by a large number of leaves in the lower layers which are poorly illuminated. Removal of such leaves may enhance photosynthesis and growth rate which was reflected in terms of seed yield (Khan and Lone, 2005).

Photosynthesis in Leaves

Leaf photosynthesis in *Brassica* species appears to be typical of most of the C_3 plants. Paul *et al.* (1990) studying photosynthetic activity in oilseed rape (*B. napus*) showed that when plants were grown at 13°C, maximal values of leaf photosynthetic rate were approximately 29 μmoles CO_2 m^{-2} s^{-1} and occurred at a temperature of 22°C. Plants grown at 30°C had lower rates of photosynthesis, while optimum rates were recorded at temperatures of approximately 31°C. Jensen *et al.* (1996) showed that oilseed rape during early growth stages, light saturated photosynthetic rate was 35-45 μmol CO_2 m^{-2} s^{-1} and stomatal resistance was 1-1.5 μmol m^{-2} s^{-1} and both varied linearly with leaf nitrogen content.

Photosynthesis and Ploidy Level

Attempts have been made to study the effect of ploidy on some attributes of photosynthesis in monogenic diploid and digenomic tetraploid species of *Brassica* since these two levels of ploidy exists in cultivated species of *Brassica*. The usual measurement of photosynthesis based on leaf area are not useful without knowing the number of cells. Expression of photosynthesis on per cell and per unit DNA has been made (Neena and Srivastava, 1994).

It has been shown of cell per unit area is lesser in allotetraploid than those of diploid (Table 5). Larger cells in polyploidy plants have higher photosynthetic capacity than smaller cells with lower chromosome number (Warner and Edward, 1988). The lesser number of cell at tetraploid level per unit area indicates a higher cellular volume. But since decrease in cell number is not as expected from the hybridization of their respective parents, it indicates a change in cell packing at tetraploid level. DNA content per unit area as well as per cell was higher in allotetraploid and their parents diploids was close to 2 (Table 5) indicating an approximate doubling of DNA content in tetraploids. Decrease in cell number was not directly proportional to increase in DNA content because polyploids employed in the study were homogenous *i.e.* the plants at two ploidy levels were evolved from different parental strain and gene action could not be expected to be strictly addative.

During plant ontogeny, in general, there was an enhancement in photosynthetic rate in the leave from early vegetative stage to opening of the first flower first anthesis in all species at both levels of ploidy (Table 6). Depression in Pn occurred at late reproductive stage *i.e.* 50% flowering when expressed on area basis. Pn at vegetative stages (both early and late) were lower in *B. oleracea* var. Botrytis (V5) and *B. oleracea* var. capitata (V4) diploids, highest in (V3) (diploid) and intermediate in rest of the species both diploid and allotetraploids.

When expressed on DNA basis, the Pn of the diploid species were either similar or higher than that of allotetraploid at all stages of growth and development. Contrarily, data showed a sharp rise in Pn of leaf tissue parallel with increase in ploidy level. Pn of V6 was almost twice as great as in V1 and V2 (primary diploid sp.) Similar observations were recorded for other allotetraploids. At reproductive stage, Pn rate could not be measured in case of V4 and V5. In rest of the diploids Pn on leaf area basis was low.

Studies on photosynthetic enzymes showed that RuBP Case activity at all the four basis of expression increased with the age of the plants in all species upto the first anthesis followed by a sharp decline (Table 7). On per cell basis, RuBP Case activity almost doubled in allotetraploids. Ploidy effects were clearly visualized when cell was used as basis of expression. Allotetraploids showed a two fold increase over their respective parental species indicating full expression of genome except in case of V7 where increase was slightly less.

Table 5 : Effect of Ploidy on Number of Cell and Content of DNA in Diploid and Allotetraploid Species of *Brassica*

Parameters	*Ploidy Level*							
	Diploids					*Allotetraploid*		
	V1	*V2*	*V3*	*V4*	*V5*	*V6*	*V7*	*V8*
No. of cells $cm^{-2} \times 10^6$	1.35	1.55	1.65	2.25	2.52	1.05	1.29	1.49
S.E. ± (× 10^6)	0.04	0.05	0.03	0.08	0.07	0.01	0.03	0.05
Content of DNA								
$\mu g\ cm^{-2}$	9.01	12.33	10.80	17.85	16.88	15.75	20.63	20.83
S.E. ±	0.07	1.08	0.95	1.23	1.55	1.33	2.01	1.88
pg $cell^{-1}$	6.66	7.95	6.55	7.93	6.69	15.01	15.98	13.98

V1 = *B. nigra;* V2 = *B. campestris* var. *Toria;* V3 = *B. campestris* var. Bron *sarson;* V4 = *B. oleracea* var. *capitata;* V5 = *B. oleracea* var. *botrytis;* V6 = *B. juncea;* V7 = *B. carinata;* V8 = *B. napus.*

Source : Neena and Srivastava (1994).

Table 6 : Effect of Ploidy on Net Photosynthesis (Pn) (CO_2 Uptake) in Diploid and Allotetraploid Species *Brassica* at Vegetative Stages

Unit of Expression	*Ploidy Level*							
	Diploids				*Allotetraploid*			
	V1	*V2*	*V3*	*V4*	*V5*	*V6*	*V7*	*V8*
Early Vegetative Stage								
µmol $m^{-2}s^{-1}$	9.47	11.49	12.63	8.21	8.85	12.37	10.10	12.25
S.E. ±	0.50	1.25	2.09	2.10	0.95	1.99	0.95	1.20
m mol (mg DNA^{-1})h^{-1}	0.38	0.34	0.42	0.17	0.19	0.28	0.18	0.21
p mol $cell^{-1}h^{-1}$	2.53	2.67	2.75	1.31	1.26	4.24	2.82	2.96
Late Vegetative State								
µmol $m^{-2}s^{-1}$	11.36	12.18	14.20	8.84	9.48	13.89	11.67	13.63
S.E. ±	0.95	1.25	2.90	1.02	0.89	1.99	0.95	1.00
m mol (mg DNA^{-1})h^{-1}	0.45	0.36	0.47	0.18	0.20	0.32	0.20	0.24
p mol $cell^{-1}h^{-1}$	3.03	2.83	3.10	1.41	1.35	4.76	3.26	3.29
First Anthesis								
µmol $m^{-2}s^{-1}$	13.88	13.26	17.68	*	*	18.31	14.52	16.41
S.E. ±	2.34	2.10	3.19	*	*	3.22	2.50	1.05
m mol (mg DNA^{-1})h^{-1}	0.55	0.39	0.59	*	*	0.42	0.25	0.28
p mol $cell^{-1}h^{-1}$	3.70	3.08	3.86	*	*	6.28	4.05	3.96

Contd...

1	2	3	4	5	6	7	8	9
			50% Flowering					
µmol $m^{-2}s^{-1}$	11.90	10.11	13.26	*	*	15.78	12.31	14.52
S.E. ±	1.00	1.35	1.25	*	*	1.91	1.31	3.22
m mol (mg DNA^{-1})h^{-1}	0.48	0.29	0.44	*	*	0.36	0.21	0.25
p mol $cell^{-1}h^{-1}$	3.20	2.35	2.89	*	*	5.41	3.44	3.51

V1 = *B. nigra;* V2 = *B. campestris* var. *Toria;* V3 = *B. campestris* var. Brown *sarson;* V4 = *B. oleracea* var. *capitata;* V5 = *B. oleracea* var. *botrytis;* V6 = *B. juncea;* V7 = *B. carinata;* V8 = *B. napus*

Source : Neena and Srivastava (1994).

Table 7 : Effect of Ploidy Level on RuBP Case Activity (CO_2 Fixed) in Diploid and Allotetraploid Species of at Vegetative Stage

Unit of Expression	*Ploidy Level*							
	Diploids				*Allotetraploid*			
	V1	*V2*	*V3*	*V4*	*V5*	*V6*	*V7*	*V8*
				Early Vegetative Stage				
S.E. ±	1.00	1.35	1.25	*	*	1.91	1.31	3.22
dpm (gFW^{-1})h^{-1} × 10^5	51.25	42.5	65.25	75.00	88.00	69.25	69.0	98.25
S.E. ± (× 10^5)	5.20	3.59	6.33	7.50	4.29	5.99	5.99	10.21
dpm cm^{-2} h^{-1} × 10^5	1.03	1.23	1.76	3.83	3.96	2.49	2.49	4.81
dpm (mg DNA^{-1})h^{-1} × 10^5	0.11	0.09	0.16	0.21	0.23	0.16	0.16	0.23
dpm $cell^{-1}$ h^{-1}	0.08	0.08	0.11	0.17	0.16	0.24	0.24	0.33

Contd....

1	2	3	4	5	6	7	8	9
Late Vegetative State								
dpm $(gFW^{-1})h^{-1} \times 10^5$	75.25	50.75	84.11	85.20	89.30	82.39	82.39	111.01
S.E. ± $(\times 10^5)$	6.39	4.99	8.12	6.79	10.23	7.31	7.31	10.32
dpm $cm^{-2}\ h^{-1} \times 10^5$	1.51	1.47	2.27	4.35	4.02	2.97	2.97	5.44
dpm $(mg\ DNA^{-1})h^{-1} \times 10^5$	0.17	0.12	0.21	0.24	0.24	0.19	0.19	0.26
dpm $cell^{-1}\ h^{-1}$	0.11	0.09	0.14	0.19	0.16	0.28	0.28	0.36
First Anthesis								
dpm $(gFW^{-1})h^{-1} \times 10^5$	105.21	89.5	124.0	*	*	124.10	124.10	131.75
S.E. ± $(\times 10^5)$	10.21	9.4	10.99	*	*	12.31	12.31	11.32
dpm $cm^{-2}\ h^{-1} \times 10^5$	2.10	2.59	3.35	*	*	4.47	4.47	6.46
dpm $(mg\ DNA^{-1})h^{-1} \times 10^5$	0.23	0.21	0.31	*	*	0.28	0.28	0.31
dpm $cell^{-1}\ h^{-1}$	0.16	0.17	0.20	*	*	0.43	0.43	0.44
50% Flowering								
dpm $(gFW^{-1})h^{-1} \times 10^5$	99.56	95.00	109.50	*	*	110.50	110.50	162.25
S.E. ± $(\times 10^5)$	7.98	9.10	8.23	*	*	11.31	11.31	15.23
dpm $cm^{-2}\ h^{-1} \times 10^5$	1.99	2.76	2.88	*	*	3.98	3.98	7.95
dpm $(mg\ DNA^{-1})h^{-1} \times 10^5$	0.22	0.22	0.27	*	*	0.25	0.25	0.38
dpm $cell^{-1}\ h^{-1}$	0.15	0.18	0.17	*	*	0.38	0.38	0.53

V1 = *B. nigra*; V2 = *B. campestris* var. *Toria*; V3 = *B. campestris* var. Bron *sarson*; V4 = *B. oleracea* var. *capitata*; V5 = *B. oleracea* var. *botrytis*; V6 = *B. juncea*; V7 = *B. carinata*; V8 = *B. napus*

Source: Neena and Srivastava (1994.)

It was reported that cellular activity of RuPB Case was higher in allotetraploids as compared to diploids. The increase was proportional to increase in chromosome complement in some cases but not in all. It is logical since allopolyploids are not derived through exclusive self duplication but rather through hybridization and polyploidy. In such cases natural selection acting on the number of size of mesophyll cells, content of chlorophyll per cell and amount of protein may result in alteration of the direct effect of doubling the number of chromosomes.

From these results it appeared that in allotetraploid *B. juncea,* the DNA content and RuBP Case activity increased two folds over its respective parents. It appears that genes are fully expressed in these allotetraploids. A corresponding increase in CER was also reported in this case at cellular level. In other two allotetraploids *B. napus* and *B. carrinata* increase in all the three parameters were reported but the values deviated from the expected sum of the values of their respective parents. The present findings indicate that *Brassica* species at tetraploid level have potential for increased photosynthesis.

Species level differences in photosynthesis and the cytoplasmic influence on inheritance has also been observed by Ramana *et al.* (2002). Photosynthetic rate in *B. nigra* and *B. carinata* is significantly higher than *B. campestris* and *B. juncea.* The former two species also had high photosynthetic rate for longer period during ontogeny.

All green surface area in the leaves, stem and pods are capable of fixing CO_2 but not equally. Chapman *et al.* (1984) reported that at early flowering, leaves were major fixing organ (67%) of total carbon fixed) with significant amount fixed by stem (30%). Despite the fact that cultivars of *Brassica* are morphologically determinate, accumulation of dry matter continued in the development of stem and reproductive organs. The late formed flowers in the apical region of raceme developed pods without any seeds or with fewer and lighter seeds, yet dry matter was invested in stem development, consequently, an extended period of flower development becomes a period of competition for dry matter utilization between stem and flowers. This implies that reduction in competition be minimized by shortening the stem size with shorter inter flower distance on different racemes (Bharghava and Tomar, 1988).

Photosynthetic assimilation and RuBP carboxylase activity declined beyond 30 days in *B. napus, B. campestris* and *B. juncea,* although total biomass accumulation in pods continued till 60-80 days after flowering suggesting thereby alternate sources for biomass accumulation during later stages of pod growth. In *B. carinata* shorter duration of pod growth coupled with relatively lower CO_2 assimilation rate and RuBP carboxylase activity accounted for lower total dry matter accumulation (Anonymous 1989).

PHOTOSYNTHESIS IN THE PODS

Photosynthetic activity of chlorophyllous reproductive structures was shown by many workers (Atkins *et al.*, 1977; Tomar *et al.*, 1979). The assimilatory activity of fruits of legumes was relatively well characterized (Lovell and Lovell, 1970; Hedley *et al.*, 1975). In oilseed *Brassica* relatively fewer studies have been made on these aspects. Allen *et al.* (1971) suggested that in rapeseed the pods themselves rather than the leaves produce the assimilates necessary for their own increase in size and weight during the pod development. Hozyo *et al.* (1972) reported that photosynthetic activity of pods of rape at early ripening stage is same as that of leaf blade but at late ripening stage pods show high rate than that of leaf blade. Inanga *et al.* (1979) reported that pods possess extensive surface area occupying the upper levels of canopy and contribute considerably to the canopy in both photosynthesis and respiration during their development. Apparent photosynthesis of the pod showed a value near to 10.5 mg $CO_2.dm^{-2}$ surface area/hr during 12-36 days after flowering and then declined rapidly as the pod matured. For the whole period of pod development the contribution of gross photosynthesis by the pod and of translocation from other oranges to total requirements were 70 ad 30% respectively.

While studying with four *Brassica* species, viz. *B. juncea, B. campestris, B. napus* and *B. carinata* Singh and Bhargava (1988) reported that net photosynthesis rate increased gradually from 10 days after flowering (DAF) to a maximum at 30 DAF and thereafter declined in all the three species except *B. carinata.* The peak of carboxylase activity corresponded to maximum photosynthetic rate. Rate of CO_2 uptake was maximum in *B. juncea* (cv. Pusa Bold), while RuBP carboxylase activity was relatively higher in *B. napus* (cv. BO-54). On the other hand pod growth duration was longer in BO-54 and *B. campestris.* The varietal differences in both the photosynthetic rate and carboxylase activity were significant. It is likely that the variation in pod yield might be explained on the basis of pods photosynthetic activity and its duration (Singh and Bhargava, 1988).

The photosynthetic contribution of fruiting structure like pod wall in building of yield in mustard has also been reported by Singal *et al.* (1987). Their studies in *Brassica campestris* L. have shown that leaf, pod wall of 'Toria' assimilate CO_2 via Calvin cycle with RuBP carboxylase catalyzing the initial carboxylation reaction and producing 3-PGA as the primary product of photosynthesis.

The rate of photosynthesis and its relation to tissue nitrogen content was studied in leaves and siliques of winter oilseed rape (*Brassica napus* L.) growing under field conditions including three rates of nitrogen application (0, 100 or 200 kg N ha^{-1} and two levels of irrigation (rainfed or irrigated at a deficit of 20 mm) (Gammelvind *et al.,* 1996). They reported that the predominant effect of increasing N application under conditions without water deficiency was enhanced expansion of photosynthetically active leaf and siliquae surfaces, while the rate of photosynthesis per unit leaf or siliquae surface area was similar in the different N treatments. Thus, oilseed rape did not increase N investment in leaf area expansion before a decline in photosynthetic rate per unit leaf area due to N deficiency could be avoided. Much less photosynthetically active radiation penetrated into high N canopies than into low N canopies. The specific leaf area increased markedly in low light conditions, causing leaves in shade to be less dense than leaves exposed to ample light. In both leaves and siliques the photosynthetic rate per unit surface area responded linearly to increasing N content upto about 2 g m^{-2}, thus showing a constant rate of net CO_2 assimilation per unit increment in N (constant photosynthetic N use efficiency). At higher tissue N contents, photosynthetic rate responded less to changes in N status. Expressed per unit N, light saturated photosynthetic rate was three times higher in leaves than in silique valves, indicating a more efficient photosynthetic N utilization in leaves than in siliques. Nevertheless, from about two weeks after completion of flowering and onwards, total net CO_2 fixation in silique valves exceeded that in leaves because siliques received much higher radiation intensities than leaves and because the leaf area declined rapidly during the reproductive phase of growth. Water deficiency in late vegetative and early reproductive growth stages reduced the photosynthetic rate in leaves and, in particular, siliques of medium and high N plants, but not of low N plants.

In a study by Mogensen *et al.* (1997) it has been shown that in *Brassica napus* cv. Global seed growth mainly depends on husk CO_2 assimilation. In irrigated plants, the net photosynthetic rate (A_{max}) was 10-13 µmol CO_2 m^{-2} s^{-1} in non-maturing pods and correlated with nitrogen content. The stomatal conductance of water vapor (gH_2O) was 0.3 mol m^{-2} s^{-1} in non-maturing pods. The photosynthetic nitrogen use efficiency (NUE) was 8.3 µmol CO_2 g^{-1} N s^{-1}, about one-third of that in leaves. The photosynthetic water use efficiency (WUE; A_{max} gH_2O^{-1}) was similar in pods and leaves. In severely drought plants, the photosynthetic rate was reduced to 38%. The seed growth rate, however, was not influenced by intermittent periods of water stress, indicating translocation of assimilates to the seeds. The drought resistant character of the pods was due to low specific area, succulence, low stomatal conductance causing a small decrease of "¨day^{-1} during soil drying and maintenance of high relative water content during severe drought.

Seed is also capable of assimilating CO_2. In seed CO_2 is incorporated into oxaloacetate via PEP-carboxylase which is rapidly converted to malate by malate dehydrogenase. Malate after decarboxylation by NADP-malic enzyme, may play a significant role in generating NADPH for biosynthesis of fatty acids. The CO_2 evolved after decarboxylation may be reduced to carbohydrate by the enzymes of the Calvin cycle. This is supported by the observation that less CO_2 is lost during light compared to dark by the isolated seed. High activity of PEP-carboxylase and higher rates of dark CO_2 fixation by seeds suggest that PEP carboxylase in seeds fixes CO_2 released during either decarboxylation of malate by malic enzyme or seed respiration (Singal *et al.* 1987).

An attempt has been made to assess the photosynthetic potential of developing seeds of oilseed rape (*Brassica napus* L.) and to compare photosynthetic properties of embryo plastids with those of leaf chloroplasts from the same species (Eastmond *et al.*, 1996). Measurements of CO_2 dependent O_2 evolution show that developing seeds of *B. napus* are photosynthetically active *in vitro.* Essentially, all of the photosynthetic activity of the developing seed is accounted for by the embryo. The rate of photosynthesis by developing embryos increased until the onset of desiccation, after which it declined, so that by maturity, embryos were no longer photosynthetically active. Photosynthetic activity was positively correlated with chlorophyll content throughout development. Comparison of the photosynthetic characteristics of leaf and embryo chloroplasts revealed that rates of uncoupled electron transport were 2.5 fold greater in those from the embryo. Light saturated rates of CO_2 dependent O_2 evolution, per unit chlorophyll, and CO_2 saturation points were similar for chloroplasts from both tissues. However, light saturation points and chlorophyll a/b ratios were lower for embryo than for leaf chloroplasts. Embryos and embryo chloroplasts also contained considerably less ribulose 1, 5-bisphosphate carboxylase/oxygenase protein per unit total protein, than leaves. Although excised embryos were capable of high rates of CO_2 dependent O_2 evolution (90-100 μmol mg^{-1} chlorophyll h^{-1}) under a saturating photosynthetic photon flux density (PPFD), low transmittance of light through the silique wall (30%), together with the high PPFD required to achieve light compensation points in developing seeds

(500 μmol m s^{-1}), suggests that photosynthesis *in vivo* is unlikely to make a net contribution to carbon economy under normal environmental conditions (Easmond *et al.*, 1996).

There are reports that indicate that light stimulates fatty acid synthesis in developing green oilseeds. For example, after removing *B. napus* seeds from siliques, the incorporation of ^{14}C from different substrates into lipids increased significantly in light (Aach and Heise, 1997) and decreased when photosynthesis was inhibited by 3-(3,4-dichlorophenyl)-1, 1-dimethylurea (Fuhrmann *et al.*, 1994). Similarly, after labeling seeds with ^{14}C-Suc, ^{14}C-acetate, or $^{3}H_2O$, Bao *et al.* (1998) reported that light enhanced *B. napus* de novo fatty acid synthesis in plastids but not the elongation of fatty acids (FA) in the cytosol. Most studies as cited above were the result of in vitro studies, using excised embryo exposed to light and therefore the situation of the seed in plant within siliques could not be assessed. Conclusion from such studies therefore may be limited because dissecting plant tissues is known to impair metabolism. Although these seeds are predominantly sink tissues, they contain chloroplasts with the thylakoid structures and enzymes of typical photosynthetic machinery. However, rather than producing carbon for export as in source tissues, photosynthesis in seeds could participate in metabolism in at least three ways. First, light reactions could produce NADPH and ATP for energetically expensive FAS; second, O_2 evolution may help prevent anoxia inside seeds (Rolletschek *et al.*, 2002); and third, the dark reactions of photosynthesis could improve the biosynthetic efficiency of seeds by refixing respiratory CO_2 and providing intermediates for metabolism.

Photosynthesis at each stage of development appears to be the main source of assimilate for current growth. Excess assimilate particularity at lower temperature when growth is slow is accumulated in shoot, root and leaves. The reserve however decreases as the pods grow and fell almost zero by end of seed fill. Reserves built up up preanthesis were only estimated to contribute about 10% directly to the final harvested yield but post anthesis condition were generally favourable for assimilate supply from current photosynthate (Bilsborrow and Norton 1984 and Norton *et al.*, 1991).

Evan (1984) showed that restructure of assimilate supply late in the vegetative phase had little effect on yield and interpreted this as the effect of mobilized reserve from stem contributing to compensatory growth under stress condition where post anthesis growth is curtailed, reserves presumably contribute a larger proportion, although seed yield may still not exceed post anthesis growth (Medham *et al.,* 1990). The number of pods and seed within them which survive the competitive stresses, assimilate shortages and the photosynthetic surfaces during flowering determine the size of the sink for photosynthates. The green areas of the pod walls, together with the amount of radiation they intercept, are then the final elements in the supply of assimilate to the seeds they contain.

Mendham (1981a) interpreted the complex supply and demand situation when comparing seed growth rates with mean daily solar radiation receipts during the seed fill period. In a year with less seeds but excellent crop growth before flowering, followed by favourable conditions for seed fill, the crop appeared 'sink limited' and individual seeds made rapid growth per day and per unit of radiation received, to a large final size. When large numbers of seeds had been retained on crops, and particularly with subsequent high temperatures and water stress, rates of seed growth per unit of radiation received were low interpreted as 'source limited'.

ABIOTIC STRESS

Amongst the abiotic stress, water, temperature and salinity are most important factors governing growth and yield of rapeseed mustard crop.

MOISTURE

Growth and yield of rapeseed mustard suffer if the water availability is limited, and the extent of damage will depend upon the variety and stage of growth and development at which the stress is experienced. Although water stress at any stage affects the growth and yield however, the most critical stage is flowering and early pod development (Richards and Thurling, 1978a), when numbers of pods and seeds are being determined.

In most of the area rapeseed mustard is grown under rainfed condition. Species level differences have been reported is drought tolerance. *B. napus* showed a better drought tolerance capacity due to higher root/shoot ratio, and greater diversion of post anthesis dry matter to seed rather than branches and pod wall. There was significant variation between and within the species (Richards, 1978), particularly *B. napus*, for a range of drought tolerance characteristics such as proline accumulation, chlorophyll stability and germination rate under water stress. Winter cultivars possessed considerable drought tolerance, probably involving the same factors as responsible for cold tolerance.

A range of techniques have been used to detect differences between cultivars in their response to water stress (Clarke and McCaig, 1982). Greater tolerance of *B. napus* to stress was shown by higher osmotic potential, lower leaf temperatures and greater uptake of $^{14}CO_2$ during stress conditions. The production of short, tuberized roots by *B. napus* under dry conditions (Potfer *et al.*, 1988) appears to be an adaptive response which can act as a reserve of water absorbing potential, and may be capable of further selection.

Water stress causes reduction in osmotic potential of leaf cell (solute concentration increased) over time to allow leaf turgor and hence transpiration and growth to be maintained as soil water potential get reduced. This is particularly important in Indian environmental condition as crops are grown on stored moisture.

In a study with *Brassica napus* and *Brassica juncea* genotypes, Naknam *et al.* (2003) have shown that genotypes with low osmotic adjustment under rainfed condition, had a yield reduction of upto 40% whereas those with high osmotic adjustment had only 0-10% yield reduction. In contrast, seed oil concentration from 41% under irrigation decreased to 38% under water deficit and the differences among genotypes were not related to osmotic adjustment. A selection criteria such as osmotic adjustment can be useful in breeding programme aiming to increase drought resistance and seed yield, especially in a water limited condition.

Because yield and drought tolerance are controlled at separate loci (Blum, 1983, Morgham, 1984), breeding for drought tolerance involves identification and transfer of morpho-physiological and biochemical traits that may impart drought tolerance to high yielding cultivars. However, limited information is available on identification and screening of such traits and genotypes for drought tolerance in Indian mustard. Clarke and Coworkers (1984) suggested that selecting for yield under dry condition should alone be more productive avenue for improvement of drought resistance until more rapid and effective screening procedures could be developed. Singh and Choudhary (2003) have suggested drought susceptibility index (based on yield reduction under drought condition). DSI as selection criteria.

Osmotic adjustment could play an important role in maintaining cell turgor potential and turgor related processes such as stomata opening, photosynthesis, shoot growth and root extension in deeper soil layers. Ashraf and Menhmood (1990) have assessed the relative drought tolerance of *Brassica* species *B. compestries, B. carinata, B. juncea* and *B. napus* and found a close relationship of biomass production and water content and water relation in response to repeated drought cycle. It has been concluded that *B. napus* was the most tolerance, *B. juncea* and *B. compestries* intermediate and *B. carinata* the most sensitive to drought stress thus emphasizing osmotic adjustment as a selection criteria for drought tolerance. The osmotic adjustment as a criteria for selection finds support also from Niknam and Turner (1999).

Singh *et al.* (2003) have shown that there is reduction in translocation of dry matter from pod wall to seed under rainfed condition and suggested as a selection criteria under limited water supply.

Kumar *et al.* (1984) showed that *B. carinata* has greater degree of osmoregulation and was able to adjust much more effectively under stress environment than *B. napus*. More wax on the leaves, stems and pods assisted the cooling process by reflecting more radiation. Growth and yield were almost double that of the *B. napus* line. Indian *B. juncea* and *B. carinata* lines showed the greatest osmoregulation and the highest yields. Canadian *B. juncea* and a range of *B. napus* lines showed the least osmoregulation and low yields, with *B. rapa* lines intermediate for both. The higher yielding lines generally use less water in the vegetative phase, allowing more for reproductive growth (Kumar *et al.* 1987).

Water use efficiency is a key indicator of productivity with a limited water supply. It is measured as dry matter produced (total or seed) per unit of water used, including both crop transpiration and soil evaporation. Water use efficiency can be used as an indirect selection criteria for seed yield in genotypic selection under abiotic stress condition (Faraji *et al.*, 2009).

B. juncea generally produces more dry matter, particularly in stem, pod wall and root system, than *B. napus,* but often a lower seed yield under adequate moisture conditions. The use of *B. carinata* and wild *Brassica* relatives with improved water use or drought tolerance characteristics is likely to be more feasible with biotechnological methods to transfer characters to cultivars with higher yield potential.

Accumulation and release of potassium by stomatal guard cells lead to changes in their turgor, resulting in stomatal opening and closing (Fischer and Hsiao, 1968). Numerous studies have shown that the application of K fertilizer mitigates the adverse effects of drought on plant growth (Sangakkara *et al.*, 2001). When plants are grown under low K supply, drought-stress induces ROS production which can be additionally enhanced, at least due to K^+ deficiency induced disturbances in stomatal opening, water relations, and photosynthesis (Marschner, 1995). The more K^+ requirement of plants under different abiotic stresses appears to be related to the inhibitory role of K^+ against reactive oxygen species (ROS) production during photosynthesis and NADPH oxidase (Cakmak, 2005).

TEMPERATURE

Temperature affects the growth and yield of mustard plant. Both lower and higher temperature at sowing time either delays or enhance the rate of germination respectively and subsequent growth and developmental processes. In Indian condition, under normal sown condition, a great part of the vegetative phase of mustard plant is completed when the temperature is relatively high but at flowering and afterward the plants are exposed to low temperature. The temperature during the entire growth period either reduces or enhances the crop duration. Thurling and Das (1980) reported that delay in sowing reduces the vegetative phase and advances the flowering time consequently reduces the total dry matter accumulation.

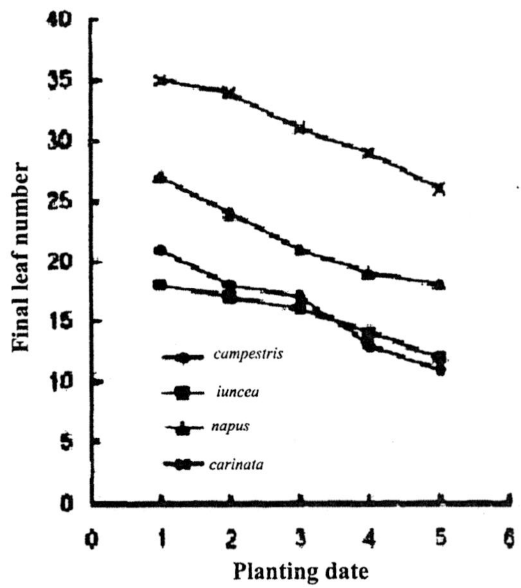

Fig. 10 : Final leaf numbers for the four *Brassica* species from each of the five planting dates (19, 29 October, 8, 28 November and 9 December, indicated as 1, 2, 3, 4 and 5 respectivly).

Source : Nanda *et al.* (1995)

In north Indian condition, the recommended sowing time is mid of October when the mean daily temperature is about 25^{0}C and maximum yields are obtained with mid October sowing. Delay in sowing reduces the final leaf number. While studying with four *Brassica* sp. Nanda *et al.* (1995) have reported that there is universal reduction in leaf number in later planting under lengthening photoperiod and lower temperature (Fig. 10)

A compression of yield of optimal sowing date (20th October) was made with the yield of seven different sowing dates from 12th October to 11 December under field condition (Bhargava and Tomar, 1991). It was observed that in all later sowings there was progressive reduction in seed yield. Results of two species viz. *B. juncea* cv. Varuna and *B. carinata* cv. are shown in table 8.

Table 8 : Effect of Delayed Planting on the Total Dry Matter (TDM) g/m^2 and Seed Yield g/m^2 on Early and Late Flowering *Brassica* Species

Species	*Parameters*	*Sowing Dates*						
		October		*November*			*December*	
		12	*22*	*1*	*11*	*21*	*1*	*11*
B. juncea (cv. Varuna)	TDM	1629	1990	895	680	628	434	298
	Seed yield	431	491	275	173	137	84	58
B. carinata (cv. Tall-1)	TDM	2000	2132	1406	1246	894	627	545
	Seed yield	393	399	283	243	195	142	128

CD at 5%: Variety = 21; Treatment = 21; Interaction = 54

Source : Bhargava and Tomar (1991)

Yield reduction each day of delay in sowing of cvs. Varuna and Tall-1 were 7.39 and 4.97 g m^{-2} d^{-1} respectively (Bhargava and Tomar, 1991). This led to conclude that for late sown conditions, extremely late flowering types with relatively high TDM at flowering *e.g. B. carinata,* yielded relatively better compared to *B. juncea* should prove profitable in obtaining higher seed yield under late sown condition.

Late sowing in mustard exposes reproductive growth to higher temperature and results in flower abortion and poor dry matter production. A considerable genotypic variation in thermotolerance particularly in the newly developed promising genotypes of mustard exists which can be used for breeding and selection of temperature tolerant high yielding mustard. Lallu *et al.* (2010) have shown a positive correlation of number of branches per plant (r=0.946) and number of siliquae per plant (r=0.982) with seed yield in late sown condition when plants at post flowering stage are exposed tò higher temperature.

There are many Brassica varieties which are resistant to high temperature so there is need to identify those termo-tolerant varieties or develop a plant with thermo-tolerant characters. Identification of genes responsible for tolerance to high or low temperature need to be done. Synthesis of polypeptides encoded by these genes as heat shock proteins (HSP) (Kaur *et al.* 2009) have been reported. Besides, plants also protects their cell and sub cellular systems from the cytotoxic effects of the reactive oxygen species (ROS) by activating antioxidative enzymes system such as catalase (CAT), peroxidase (POX), superoxide dismutase (SOD), glutathione reductase (GR), etc. Not much information is available on the involvement of these anti-oxidative enzymes in temperature tolerance in Indian mustard.

SALT

Most of the *Brassica* species have been categorized as moderately salt tolerant, with the amphidiploid species being the relatively salt tolerant in comparison with the diploid species. Due to the higher salt tolerance of the amphidiploids, it has been suggested that their salt tolerance has been acquired from the A (*B. campestris*) and C (*B. oleracea* L.) genomes. However, significant inter and intra-specific variation for salt tolerance exists within brassicas, which can be exploited through selection and breeding for enhancing salt tolerance of the crops. There are contrasting reports regarding the response of these species to salinity at different plant developmental stages, but in most of them it is evident that they maintain their degree of salt tolerance consistently throughout the plant ontogeny. The pattern of uptake and accumulation of toxic ions (Na^+ and Cl^-), in tissues of plants subjected to saline conditions appears to be mostly due to mechanism of partial ion exclusion (exclusion of Na^+ and/or Cl^-) in most of the species, although ion inclusion in some cases at intra-specific levels has also been observed. Maintenance of high tissue K^+/Na^+ and Ca^{2+}/Na^+ ratios has been suggested as an important selection criterion for salt tolerance in brassicas. Osmotic adjustment has also been reported in Brassica plants subjected to saline conditions, but particularly to a large extent in salt tolerant species or cultivars. The roles of important organic osmotica such as total soluble sugars, free amino acids, and free proline (Kumar 1984), which are central to osmotic adjustment, have been largely discussed. Madan *et al.* (1995) have reported that the activities of proline biosynthetic enzymes pyrroline 5 carboxylate reductase and ornithin aminotransferase increased under salt stress both in seedling and leaves of *Brassica juncea* L. While the activity of proline degrading enzyme proline oxidase decreased under salt stress in both the tissue. Thus the trend of change in the enzyme activities was in tune with increase in proline level.

Salt Stress Environment

Proline synthesizing enzymes		Proline degrading enzyme Proline oxidae
Pyrroline 5 carboxylate reducatase Ornithine aminotransferase	↑ increase	↓ decrease

PHYTOREMEDIATION

In recent years, there has been considerable interest in the use of plants for remediation of environmental contamination particularly heavy metals contamination of soil (Phyto-remediation). *Brassica juncea* is a dry land plant which has a root system that is known to hyperaccumulate certain heavy metals. This species can take up significant quantities of Pb and Cd under a variety of experimental conditions (Dushenkov *et al.*, 1995; Kumar *et al.*, 1995; Salt *et al.*, 1998; Vassil *et al.*, 1998; Jiaug *et al.*, 2000), including rhizofiltratin of toxic ions from water that contains a variety of other non toxic ionic ions (Salt *et al.*, 1998). In addition, it has also been demonstrated in *B. juncea* that Cd tolerance and accumulation is enhanced by over expression of an enzyme that catalyses the synthesis of metal chelating peptides (Zhu *et al.*, 1999a, b).

Transgenic Indian mustard (*Brassica juncea* L. Czern.) plants overproducing the enzymes gamma-glutamylcysteine synthetase (ECS) or glutathione synthetase (GS) were shown to have increased levels of the metal binding thiol peptides phytochelatins and glutathione, and enhanced Cd tolerance and accumulation. Furthermore, transgenic Indian mustard plants overexpressing adenosine triphosphate sulfurylase (APS) were shown to have higher levels of glutathione and total thiols and were able to accumulate Se in their leaves (Bennett *et al.*, 2003, Banuelos *et al.*, 2005). Genetically engineered plants of mustard can perform phytoremediation successfully under field conditions.

Indian mustard (*Brassica juncea*) has been shown to be effective at accumulating high tissue concentrations of lead when grown in contaminated soil with the addition of a chelating agent, such as EDTA. Copper removal from the soil by *Brassica juncea* (L.) has been reported by Ariyakanon and Winaipanich (2006).

Besides removing heavy metals, *Brassica napus* hairy root which produce peroxidases, enzyme involved in biodegration processes that could be used for removal of 2,4-dichlorophenon (2,4-DCP) a common contaminants in industrial effluents that is highly toxic to human and aquatic life. Agostini *et al.* (2003) have shown that roots were able to remove 2,4-DCP from aqueous solutions containing 100-1000 mg/l in the presence of H_2O_2 concentrations ranging from 5 to 10 mM. After a short period of incubation (15 min.), high removal efficiencies were achieved (91-94%) and maximal removal, of approx. 97-98%, was obtained with 1 h of reaction. High removal efficiencies (93-95%) were observed in a broad pH range (pH 3-9), reaching 98-99% in the range pH 4-8. Moreover, roots could be re-used, almost for six consecutive cycles, to remove 2,4-DCP. The oxidation catalysed by peroxidases would be the main mechanism involved in this process. The results suggest that these cultures could be useful tools for phytoremediation.

CONCEPT OF A PLANT TYPE

Development of the ideotype concept has focused the

attention of plant physiologists on identification of simple morphological characters which have some influence on physiological processes determining the yield of the economic organs. Characters such as leaf inclination and leaf shape, for example, are often simply inherited and can greatly influence crop canopy structure and radiation interception. Such characters could be rapidly modified by selection to increase crop photosynthesis and yield. Ideotype definitions integrate information on these relationships and provide plant breeders with a clear blueprint of the characteristics of a high yielding cultivar in a specified environment.

Fig. 11 : Plant ideotype for crop *Brassicas* for monoculture conditions where inputs are non-limiting. Figures on the bars in the leaf axil represents number of pods on the branches.

(*Source :* Bhargava and Tomar, 1988)

Research on oilseed *Brassica* species (Thurling, 1991) has identified a number of highly heritable characters which provide a basic framework for the definition of a yield ideotype for spring rape (*B. napus*) in western Australia. These include an optimal time of flowering, apetalous flowers and long upright pods to improve light penetration of the dense pod canopy, and pod shattering resistance to allow for direct harvesting of the crop. A major advantage in breeding for this ideotype is that genes for certain characters can be easily introgressed from the related species *B. campestris* and *B. juncea.*

Bhargava and Tomar (1988) proposed a plant type of *Brassica* which may yield 5 t/ha. Plant type was to have following characteristics (Fig. 11) : plant height 1-1.25 m, only 5-6 primary branches at an upper plant profile, the main racemes bearing only 40 pods whereas the lower branches only 15 each and the upper two branches only 20 each, plants to have smaller and thicker leaves with higher photosynthetic rate and low dark respiration, each pod having 20 seeds and each seed should weigh 5 mg. There could be 7-10 leaves on the main stem to support carbohydrate and nitrogen requirement for the required seed growth. This plant in isolation would be a poor yielder but should perform well in higher population with non-limiting inputs.

However, for condition particularly for late sown condition, Kumar and Srivastava (2003) have postulated that a genotype having impressive vegetative structure, *i.e.* taller in height and higher number of primary branches associated with modest seed test weight, deep green leaves having high chlorophyll content and lower range of proline content would be ideal for late sown Indian mustard.

REFERENCES

Agostini, E., Coniglio, M.S., Milrad, S.R., Tigier, H.A. and Giulietti, A.M. (2003). Phytoremediation of 2,4-dichlorophenol by *Brassica napus* hairy root cultures. *Biotechnol. Appl. Biochem.*, **37**(2) : 139-144.

Alemaheyu, N. and Becker, H. (2002). Genotypic diversity and patterns of variation in germplasm material of Ethiopian mustard (*Brassica carinata* A. Braun). *Genetic Resources and Crop Evolution*, **49**(6) : 573-582.

Almond, J.A., Dawkins, T.C.K., Done, C.J. and Ivins, J.D. (1984). Cultivations for winter oilseed rape (*Brassica napus* L.). *Aspects of Applied Biology*, **6** : 67-79.

Alonso, L.C., Fernandez-Serrano, O. and Fernandez-Escobar, J. (1991). The outset of a new oilseed crop: *Brasscia carinata* with a low crucic acid content. In : McGregor, D.I. (ed.) *Proceedings of the Eighth International Rapeseed Congress*, Saskatoon, Canada, Organizing Committee, Saskatoon, pp. 170-176.

Anonymus (1991). Plant physiology research at IARI. Pp. 177

Anonymous (1989). Research Highlights (1977-1988), Division of Plant Physiology, IARI, New Delhi p. 60.

Anonymous (2006) Souvenir Four decades of Division of Plant Physiology (966-2006). Indian Agricultural Research Institute, New Delhi-12, pp. 115.

Ariyakanon, N. and Winaipanich, B. (2006). Phytoremediation of copper contaminated soil by *Brassica juncea* (L.) Czern and Bidens alba (L.) DC Var. *J. Sci. Res. Chula. Univ.*, **31**(1) : 49.

Ashraf, M. and Mehmood, S. (1990). Response of four *Brassica* thesis to drought stress. *Environmental and Experimental Botany*, **30**(1) : 93-100.

Ashraf, M. and Sharf, R. (1998). Assessment of inter cultivar/line variation of drought resistance in a potential oilseed crop. Ethiopian mustard (*Brassica carinata* A. Br.). *Achieves of Agronomy and Soil Science*, **43**(4) : 251-265.

Bahl, P.N. and Jain, H.R. (1977). Association among agronomic characters and plant ideotype in chickpea (*Cicer arietinum* L.). *Pflanznzuechtung*, **79** : 154-159.

Banuelos, G., Terry, N., Leduc, D.L., Pilon-Smits, E.A.H. and Mackey, B. (2005). Field trial of transgenic Indian mustard plants shows enhanced phyto-remediation of selenium-contaminated sediment. *Environ. Sci. Technol.*, **39** : 1771-1777.

Bao, X.M., Pollard, M. and Ohlrogge, J. (1998). The biosynthesis of erucic acid in developing embryos of *Brassica rapa*. *Plant Physiol.*, **118**: 183-190.

Bennett, L.E., Burkhead, J.L., Hale, K.L., Terry, N., Pilor, M. and Pilor-Smits, E.A.H. (2003). Analysis of transgenic Indian mustard plants for phyto-remediation of metal-contraindicated mine tailings. *J. Environ. Qual.*, **32** : 432-440.

Bhargava, S.C. and Tomar, D.P.S. (1990). Physiological contribution for increased yield in rapeseed mustard. In: Sinha, S.K. *et al.* (ed). *Proc. International Cong. Pl. Physiol.*, 1988, Delhi, **1** : 352-360.

Bilsborrow, P.E. and Norton, G. (1984). A consideration of factors affecting the yield of oilseed rape. *Aspects of Applied Biology*, **6** : 91-99.

Blum, A. (1983). Evidence for genetic variability in drought resistance and its implication for plant breeding. *In:* Drought resistance in crops with emphasis on rice. IRRI, Los Banos, Philippines, 53-68.

Bunting, E.S. (1986). Oilseed rape in perspective. In: Scarisbrick, D.H. and Daniels, R.W. (eds.). *Oilseed Rape*. Collins, London, pp. 1-31.

Cakmak, I. (2005). K alleviates detrimental effects of abiotic stresses in plants. *J. Plant Nutr.,* **168** : 521-530.

Chapman, J.F., Daniels, R.W. and Scarisbrick, D.H. (1984). Field studies on ^{14}C. Assimilate fixation and movement in oil-seed rape (*B. napus*). *Journal of Agricultural Science, Cambridge,* **102** : 23-31.

Chitra Devi, L., Kant, K. and Dadlani, M. (2003). Effect of size granding and ageing on sinapine leakage, electrical percentage in the seed of mustard (*Brassica juncea* L.). *Seed Sci. Technol.,* **31** : 505-509.

Clarke, J.M. and McCaig, T.N. (1982). Leaf diffusive resistance, surface temperature, osmotic potential and $^{14}CO_2$ assimilation capability as indicators of drought intensity in rape. *Canadian Journal of Plant Science,* **67** : 785-789.

Dushenkov, V., Kumar, P.B.A.N., Motto, H. and Raskin, I. (1995). Rhizofiltration : the use of plants to remove heavy metals from aqueous streams. *Environ. Sci. & Tech.,* **29** : 1239-1245.

Elliott, R.H., Mann, L.W. and Olfert, O.O. (2007). Effects of seed size and seed weight on seedling establishment, seedling vigour and tolerance of summer turnip rape (*Brassica rapa*) to flea beetles, *Phyllotreta* spp. *Canadian Journal of Plant Sci.,* **87**(2) : 385-393.

Evans, E.J. (1984). Pre-anthesis growth and its influence on seed yield in winter oilseed rape. *Aspects of Applied Biology,* **6** : 81-90.

Eyidogan, F. and Tufan, O.Z.M. (2006). Effect of salinity on antioxidant response of chickpea seedlings. *Acta Physiologiae Plantrum,* **29**(5) : 485-493.

Fiaroji, A., Latifi, N., Soltani, A. and Rad, A.H.S. (2009). Seed yield and water use efficiency of canola (*Brassica napus* L.) as affected by high temperature stress and supplemental irrigation. *Agricultural Water Management,* **96**(1) : 132-140.

Fuhrmann, J., Johnen, T. and Heise, K.P. (1994). Compartmentation of fatty acid metabolism in zygotic rape embryos. *J. Plant Physiol.,* **143** : 565-569.

Gammelvind, L.H., Schjoerring, J.K., Mogensen, V.O., Jensen, C.R. and Bock, J.O. (1996). Photosynthesis in leaves and siliques of winter oilseed rape (*Brassica napus* L.). *Plant and Soil,* **186**(2) : 227-236.

Getinet, A., Rakow, G., Raney, J.P. and Downey, R.K. (1997). Glucosinolate content in interspecific cross of *Brassica carinata* with *B. juncea* and *B. napus*. Plant Breeding, **116**(1) : 39-46.

Hegde, J.C. (1976). A systematic and geographical survey of the world cruciferеae. In : The Biology and Chemist. J.G., A.J. McLeod and B.M.G. Jones (Eds.). Academic Press, New York, pp. 1-45.

Kaur, P. Ghai, N. and Sangha, M.K. (2009). Induction of thermo tolerance through heat acclimation and salicylic acid in *Brassica* species. *African J. Biotechnology*, **8**(4) : 619-625.

Khan, N.A. and Lone, P.M. (2005). Effects of early and late season defoliation on photosynthesis, growth and yield of mustard (*Brassica juncea* L.). *Braz. J. Pl. Physiol.*, **17** : 181-186.

Kumar, A., Singh, D.P. and Singh, P. (1987). Genotypic variation in the responses of *Brassica speices* to water deficit. *Journal of Agricultural Science, Cambridge*, **109** : 615-618.

Kumar, A., Singh, P., Singh, D.P., Singh, H. and Sharma, H.C. (1984). Differences in osmoregulation in *Brassica* species. *Annls of Botany*, **54** : 537-541.

Kumar, D. (1984). The value of certain plant parameters as an index for salt tolerance in Indian mustard (*Brassica juncea* L.). *Plant and Soil*, **79** : 261-272.

Kumar, N. and Srivastava, S. (2003). Plant ideotype of Indian mustard (*Brassica juncea* L.) for late sown condition. *Indian J. Genet.*, **63**(4) : 355.

Kumar, P.B.A.N., Dushenkov, V., Motto, H. and Raskin, I. (1995a). Phytoextraction : The use of plants to remove heavy metals from soils. *Environ. Sci. and Tech.*, **29**(5) : 1232-1238.

Kumar, P.B.A.N., Dushenkov, V., Motto, H. and Raskin, I. (1995b). Rhizofiltration: The use of plants to remove heavy metals from aqueous streams. *Environ. Sci. and Tech.*, **29**(5) : 1239-1245.

Leterme, P.)1988b). Modelisation due fonctionnement du peuplement de colza d'hiver en fin de cycle: elaboration des composantes finales ldu rendement. In: *Colza: Physiologie et Elaboration du Rendement* CETIOM Paris pp. 124-129.

Lim, J.M., Salido, A.L. and Butcher, D.J. (2003). Phytoremediation of lead using Indian mustard *Brassica juncea* L. with EDTA and electrodies. *Microchemical Journal*, **76** : 3-9.

Love, H.K., Rakow, G., Raney, J.P. and Downey, R.K. (1991). Breeding improvements towards canola quality *Brassica juncea*. In: McGregor, D.I. (ed.) *Proceedings of the Eight International Rapeseed Congress*, Saskatoon, Canada, Organizing Committee, Saskatoon, pp. 164-169.

Madan, S., Nainawatee, H.S., Jain, R.K. and Chowdhury, J. (1995). Proline and proline metabolizing enzymes in *in vitro* selected NaCl-tolerant *Brassica juncea* L. under salt stress. *Annals of Botany,* **76**(1) : 51-57.

Major, D.J. (1977). Influence of seed size on yield and yield components of rape. *Agronomy Journal,* **69** : 541-543.

Major, D.J., Bole, J.B. and CHarnetski, W.A. (1978). Distribution of photosynthates after $^{14}CO_2$ assimilation by stems, leaves and pods of rape plans. *Canadian Journal of Plant Science,* **58** : 783-787.

Malik, R.S. (1990). Prospects for *Brassica carinata* as an oilseed crop in India. *Experimental Agriculture,* **26** : 125-129.

Marschner, H. (1995). Mineral nutrition of higher plants. Academic Press San Diego, USA

Mendham, N.J. and Scott, R.K. (1975). The limiting effect of plant size at inflorescence initiation on subsequent growth and yield of oilseed rape (*Brassica napus*). *Journal of Agricultural Science, Cambridge,* **84** : 487-502.

Mendham, N.J., Russell, J. and Buzza, G.C. (1984). The contribution of seed survival to yield in new Australian cultivars of oil-seed rape (*Brassica napus*). *Journal of Agricultural Science, Cambridge,* **103** : 303-316.

Mendham, N.J., Russell, J. and Jarosz, N.K. (1990). Response to sowing time of three contrasting Australian cultivars of oilseed rape (*Brassica napus*). *Journal of Agricultural Science, Cambridge,* **114** : 275-283.

Mendham, N.J., Shipway, P.A. and Scott, R.K. (1981a). The effect of delayed sowing and weather on growth, development and yield of winter oil-seed rape (*Brassica napus*). *Journal of Agricultural Science, Cambridge,* **96** : 389-416.

Mendham, N.J., Shipway, P.A. and Scott, R.K. (1981b). The effects of seed size, autumn nitrogen and plant population density on the response to delayed sowing in winter oil seed rape (*Brassica napus*). *Journal of Agricultural Science, Cambridge,* **96** : 417-428.

Mogensen, V.O., Jensen, C.R., Mortensen, G., Andersen, M.N., Schjoerring, J.K., Thage, J.H. and Koribidis, J. (1997). Pod photosynthesis and drought adaptation of field grown rape (*Bassica napus* L.). *European J. Agron.,* **6** : 295-307.

Morgan, J.M. (1984). Osmoregulation and water stress in higher plants. *Annals of Plant Physiol.,* **35** : 292-319.

Morinaga, T. (1934). On the chromosome number of *Brassica juncea* and *Brassica napus,* on the hybrid between the two and on offspring time of the hybrid. *Jap. J. Genet. (Japanees),* **9** : 161-163.

Nanda, R., Bhargava, S.C. and Rowson, H.M. (1995). Effect of sowing date on rates of leaf appearance, final leaf number and areas in *Brassica campestris, B. juncea, B. napus* and *B. carinata. Field Crop Res.,* **42** : 125-134.

Neena and Srivastava, G.C. (1994). Effect of ploily on photosynthesis in Brassica species. *Indian J. Pl. Physiol.,* **37**(4) : 237-241.

Niknam, S.R. and Turner, D.W. (1999). Physiological aspects of drought tolerance in *Brassica napus* and *B. juncea.* 10th International Rapeseed Congress, Canberra, Australia.

Norton, G., Bilsborrow, P.E. and Shipway, P.A. (1991). Comparative physiology of divergent types of winter rapeseed. *In:* McGregor, D.I. (ed.) *Proceedings of the Eighth International Rapeseed Congress,* Saskatoon, Canada. Organizing Committee, Saskatoon, pp. 578-582.

Oram, R.N. and Kirk, J.T.O. (1992). Breeding Indian mustard for Australian conditions. In: Hutchinson, K.J. and Vickery, P.J. (eds.) *Proceedings of the Sixth Australian Agronomy Conference,* Australian Society of Agronomy, Armidale, New South Wales, pp. 467-470.

Peter Eastmond, Lucie, Kolaca and Stephen Rawsthorne (1996). Photosynthesis by developing embryos of oilseed rape (*Brassica napus* L.). *J. Experimental Botany,* **47**(11) : 1763-1769.

Potfer, J.P., Merrien, A. and Vartanian, N. (1988). Etude *in situ* due systeme racinaire due colza de printemps en condition de secheresse. In: *Colza: Physiologie et Elaboralion due Rendement,* CETIOM, Paris, pp. 47-53.

Prakash, S. (1980). Cruciferous oilseeds in India. In: In *Brassica* Crops and wild allies. Biology and breeding, Trsunoda S., Hinata, K. and Gomez-Campo (Eds.) *Japan Scientific Society Press Tokyo,* pp. 151-163.

Prakash, S. and Hinata, K. (1980). Taxonomy, cytogenetics and origin of crop Brassica, a review. *Opera Botanica,* **55** : 1-57.

Pravish, O. (1961). Food and drinks in ancient India. Munshi Ram Manohar Lal, Delhi, pp. 165-168.

Ramana, S., Sharma-Natu, P. and Ghildiyal, M.C. (2002). Cytoplasmic effect on photosynthesis and ribulose-1-5 bisphosphate carboxylase activity in *Brassica* species. *Euphytica,* **123**(3) : 361-365.

Richards, R.A. (1978). Variation within and between species of rapeseed (*Brassica campestris* and *B. napus*) in response to drought stress. III. Physiological and physicochemical characters. *Australian Journal of Agricultural Research,* **29** : 491-501.

Rischer, R.A. and Hsiao, T.C. (1968). Stomatal opening in isolated epidermal strips of *Vicia faba.* II. Response to KCl concentrations and the role of potassium absorption. *Plant Physiol.,* **43** : 1955-1958.

Rolletschek, H., Borisjuk, L., Koschorreck, M., Wobus, U. and Weber, H. (2002). Legume embryos develop in a hypoxic environment. *J. Exp. Bot.*, **53** : 1099-1107.

Salt, D.E., Smith, R.D. and Raskin, I. (1998). Phytoremediation. *Ann. Rev. Plant Physiol and Molecular Biol.*, **49** : 643-668.

Sandhu, J.S., Brar, H.S. and Verma, M.M. (1990). Inheritance of a plant type mutant and its utilization in chickpea. *Euphytica*, **48** : 111-112.

Sangakkara, U.R., Frehner, M. and Nosberger, J. (2001). Influence of soil moisture and fertilizer potassium on the vegetative growth of mungbean (*Vigna radiate* L. Wilczek) and cowpea (*Vigna unguiculata* L. Walp.). *J. Agron. Crop Sci.*, **186** : 73-81.

Sedgley, R.H., Siddique, K.H.M. and Walton, G.H. (1990). Chickpea ideotypes for Mediterranean environments. In: van Rheenen, H.., Saxena, M.C., Walby, B.J., Hall, S.D. (eds.), Chickpea in the nineties: proceedings for the second international workshop on chickpea improvement. ICRI-SAT, Patancheru, AP, India, pp. 87-91.

Shantakumari, P. and Sinha, S.K. (1972). Variation I chlorophylls and photosynthetic rates of cultivars of Bengal gram (*Cicer arietinum* L.). *Indian J. Exp. Biol.*, **13**(2) : 208-209.

Singal, H.R., Sheoran, I.S. and Singh, R. (1987). Photosynthetic carbon fixation characteristics of fruiting structures of *Brassica campestris* L. *Plant Physiol.*, **83** : 1043-1047.

.Singh, N. and Bhargava, S.C. (1988). Pod wall photosynthesis of *Brassica* species. In: Proc. of the International Congress of Plant Physiology, 1988, New Delhi Pub. Int. Soc. Pl. Physiology and Biochemistry, Vol. 1, pp. 386-388.

Singh, S.P. and Choudhary, A.K. (2003). Selection criteria for drought tolerance in Indian mustard (*Brassica juncea* L. Czern & Coss.). *Indian J. Genet.*, **63**(3) : 263-264.

Smith, L.J. and Scarisbrick, D.H. (1990). Reproductive development in oilseed rape (*Brassica napus* cv. Bienvenu). *Annals of Botany*, **65** : 205-212.

Srivastava, J.P. and Balkrishna (2003). Environmental parameters influencing phonological development of mustard in relation to yield. *Indian J. Pl. Physiol.*, **8** : 349-353.

Thurling, N. (1991). Application of the ideotype concept in breeding for higher yield in the oil seed brassica. *Field Crop Res.*, **26** : 201-219.

Thurling, N. and Das, L.D.V. (1984). The relationship between pre-anthesis development and seed yield of spring rape (*Brassica napus*). *Aust. J. Agric. Res.*, **31** : 25-36.

Thurling, N. and Das, L.D.V. (1984). The relationship between pre anthesis development and seed yield of spring rape (*Brassica napus* L.). *Aust. J. Agric. Res.*, **31** : 25-36.

Torsell, B. (1959). Hardness and survival of winter rape and winter turnip rape. Department of Plant Husbandry (Crop production). Royal School of Agriculture, Sweden. Publication No. 15.

U.N. (1935). Genome analysis in *Brassica* with special refence to the experimental formation of *B. napus* and peculiar mode of fertilization. *Japanese, J. Bot.*, **7** : 389-452.

Van Huysen, T., Terry, N. and Pilon-Smits, E.A. (2004). Exploring the selenium phytoremediation potential of transgenic Indian mustard overexpressing ATP sulfurylase or cystathionine-gamma-synthase. *Int. J. Phytoremediation*, **6**(2) : 111-118.

Vassil, A.D. (1998). The role of EDTA in lead transport and accumulation by India mustard. *Plant Physiol.*, **1178** : 44-85

Vavilov, N.I. (1949). Origin, variation, immunity and breeding of cultivated plants. *Chronica Botanica*, **13** : 15-19.

Warner, D.A. and Edward, G.E. (1988). C_4 photosynthesis and leaf amatory in diploid and autotetraphoid penniset in americanum (pearl millet). *Plant Sci.*, **56** : 85-92.

Wiknam, S.R., Ma, Q. and Turner, D.W. (2003). Osmotic adjustment and seed yield of *Brassica napus* and *B. juncea* genotypes in a water limited environment in south western Australia. *Aust. J. Expt. Agric.*, **43** : 1127-1135.

Yan, Z. (1990). Overview of rapeseed production and research in China. Proc. Int. Canola Conf. Potash and Phosphate Institute. *Atlanta.*, 29-35.

Zhu, Y.L., Pilon-Smits, E.A., Tarun, A.S., Weber, S.U.,Jouanin, L. and Terry, N. (1999a). Cadmium tolerance and accumulation in Indian mustard is enhanced by over expressing gamma-glutamylcysteine synthetase. *Plant Physiol.*, **121** : 1169-1178.

Zhu, Y.L., Zayed, A.M., Quian, J.H., De-Souza, M. and Terry, N. (1999b). Phytoaccumulation of trace elements by wetland plants. II. Water hyacinth. *J. Environ. Quality*, **28** : 339-344.